INTELLIGENT TESTING, CONTROL AND DECISION-MAKING FOR SPACE LAUNCH

INTELLIGENT TESTING, CONTROL AND DECISION-MAKING FOR SPACE LAUNCH

Yi Chai
Shangfu Li

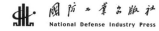

National Defense Industry Press

Library of Congress Cataloging-in-Publication data applied for.

ISBN: 9781118889985

A catalogue record for this book is available from the British Library.

Set in 10/12pt Times by SPi Global, Pondicherry, India
Printed and bound in Singapore by Markono Print Media Pte Ltd

1   2015

Contents

Preface

As an important part in astronautical engineering, testing and control for space launch comprises several systems, such as testing, tracking, telemetry, and command (TT&C), communication, meteorology, and service support, and is used for testing, launching, tracking measurement, and safety control of spacecrafts and launch vehicles. In terms of the practical problems in space testing and launch application, this book has a deep insight into theoretical research and technology application.

With a focus on informationalization, automation, and intelligentization, this book discusses the newest technologies of testing and control for space launch; it elaborates the development history and technological features of space testing and control systems and also makes a detailed analysis and summary on data processing, fault diagnosis, safety control, and decision-making. Book contents include networks of testing and control, intelligent analysis on testing parameters, intelligent fault diagnosis, safety control modeling, intelligent decision-making, and the application in space launch. On the aspect of space testing and data processing, the book describes network architecture featured by information sharing and interaction, studies the intelligent analysis and processing of testing data, and researches intelligent fault diagnosis of launch system. On the safety control of space launch and flight, the book studies safety control modeling based on telemetry and tracking data fusion, fault diagnosis, and safety assessment, which includes dropping point calculation of launch vehicles, debris spreading area, and poison gas leak model and provides support for analysis and decision-making of safety control.

Being theoretical and practical, the book can be regarded as a valuable reference for scientists and engineers in space testing and launch field. On the basis of networking and intelligentization, the book focuses on improving key technologies and theories on testing and control, fault diagnosis, and decision-making. In conclusion, the book plays a significant and positive role in space launch.

Sun Jiadong
Academician of Chinese Academy of Sciences
Chief designer of China's Lunar Exploration Project

Introduction

Testing and control for space launch is always an important part in aerospace studies and is one of the most fundamental and imperative parts in astronautic engineering. The aim of the subject is to fulfill optimal control and decision-making during the space launch by analyzing key statistics in the process of space launch and combining systematic information of testing, tracking, telemetery and command (TT&C), communication, meteorology, and service support.

Testing and control technology for space launch was initially originated from guided missile science. Without an independent testing and control system, early launch technology of guided missiles was very basic and relied mainly on manual testing. With the development of space technology and carrier rockets, launch testing techniques experienced great progress. All advanced countries in space studies, including the United States and the Soviet Union as two pioneers, spared no efforts in the study of testing and control technology for space launch. Gradually, the testing and control system for space launch became a complicated system composed of several subsystems such as testing, TT&C, communication, meteorology, and service support.

It was in 1956 when China began developing its own space programs. Similar to other countries, China also experienced three stages in the development of its testing and control system: manual testing, electromechanical testing, and computer automatic testing and control.

The development of information technology and high-tech applications in the twenty-first century played a significant role in expanding the database and multiplying data types for space launch. All the data obtained should be analyzed both in a separate and comprehensive way, thus adding to the complexity of system diagnosis; all types of data should be optimized and integrated, thus adding to the complexity of comprehensive assessment. Accordingly, a higher standard was set for launch testing and control system. It is required to collect data in real-time and display data in multiple ways, to assess and optimize data in a comprehensive manner, to make accurate judgments, and to achieve intelligent decision-making.

That said, computer testing and control systems for space launch cannot meet every afore-mentioned requirement, which shows that integration technology and intelligent technology must be an inevitable trends in future studies of testing and control systems for space launch.

Intelligent testing, control, and decision-making technology imply obtaining all information automatically and realizing testing, control, diagnosis, monitoring, and decision-making during space launch on the basis of computer science, communication, control, operational research, real-time modeling, artificial intelligence, and expert systems. Just as the name indicates, the significant feature of the technology is "intelligent," meaning that it possesses the analytic and decision-making ability to solve specific problems. The whole system consists of parameter testing, network transmission and control, comprehensive data possessing and analysis, condition monitoring and trend analysis, fault diagnosis, and intelligent decision-making.

Intelligent testing, control, and decision-making for space launch are challenging domains that hold great significance for applying intelligence theories and technology to space launch testing and control. They have played an important part in boosting the space sector and increasing launch efficiency and success rate. Making great progress in these domains is a common goal for launch centers all over the world.

The main purpose of this book is testing, control, and decision-making for space launch on the basis of computer science, communication and control technology, artificial intelligence, intelligent information processing technology, intelligent fault diagnosis, and data fusion. Having considered the demand of testing and control for space launch and studied the practical situation in launch centers, the authors make a systematic analysis and illustration of intelligent testing, control, and decision-making techniques and technology for space launch. This book consists of six chapters.

Chapter 1 is an overview of current status, characteristics, and existing problems in the present space testing and control system. It also introduces the historical development of this system and elaborates its related concepts, system components, and key technology.

Then Chapter 2 introduces the bus network and overall command network for testing, launching, and control. It covers the network architecture for automatic testing and launching as well as long-distance control network architecture for space launch. In addition, this chapter also discusses the key techniques of bus networks and presents cases and examples of network construction respectively based on 1553B, LXI, and field bus, ending up with the overall architecture and design of networks of command systems for space testing, launching, and control.

Chapter 3 investigates the intelligent analysis and processing for testing data. Focusing on the features of time varying, multiscale, nonlinearity, and dynamic that testing data possess, this chapter introduces the intelligent analysis and processing methods for testing data on the basis of wavelet, cluster, rough set, and principal component analysis. Also, this chapter discusses anti-noise measures, singular point detection, consistency analysis, simplification, and correlation analysis for testing data in collection and transmission.

Chapter 4 is about intelligent fault diagnosis for space testing and launching. The method of multidimensional scaling principal component analysis (MSPCA) based on data driving is applied in fault diagnosis from the angle of feature sample extraction. On the basis of graph theory models, the ant colony algorithm, and neural networks, the author discusses electric leakage of rockets, the intelligent diagnosis model for sneak path faults, and fault diagnosis for simulation circuits.

Chapter 5 proposes the space launch flight safety control model and intelligent decision; discusses the intelligent decision of safety control; points out the effects of flight trajectory, speed, attitude, and other key factors of safety control; establishes calculation models for safety control parameters like flight trajectory fusion, dropping point calculation, and safety channel calculation; and states the rocket explosion mode by liquid propellant fault under fault state, shrapnel spreading model, poison gas leak model, and the knowledge representation model of safety control. It also simulates safe rocket flight, analyzes the site situation, and introduces intelligent control decisions and emergency responses based on the dropping point, the debris dispersion of an exploded rocket, and concentration of toxic gas leaks.

Chapter 6 draws a vision of future aerospace technologies and methods of flight test, launch, and control; discusses the demands for research on new types of testing techniques and theories, fault diagnosis and forecast, real-time control and decision-making in the development of aerospace; and predicts the trends of informatization, intellectualization, and integration of the aerospace test, and launch and control system.

Yi Chai
Shangfu Li

1

Overview of Testing and Control for Space Launch

Astronautical engineering is recognized as a scientific domain that exerted great impact on human society in the twentieth century. Human space activities have stimulated human imagination and innovation. Space launch is, without any doubt, a systematic engineering because of its difficulty, complexity, high reliability, and high risk. With further human exploration of space resource in the twenty-first century, space entry and research are supposed to be more economic, safe, and fast, which requires space launch to be featured by high reliability and accuracy.

This chapter is an overview of testing and control for space launch. The author intends to make a brief introduction of research subject, basic content, and related concepts. By introducing different development stages of spacecrafts and space launch vehicles, the author elaborates the features of testing and control system in different space development stages; focusing on intelligent data collection, processing, analysis, control and decision-making, the author discusses online detection, distributed processing, diagnosis, and decision-making control in the process of space launch. Readers are expected to have a general picture of technology and theories on testing and control for space launch.

1.1 Survey of Space Launch Engineering

Astronautical engineering refers to a comprehensive system involving exploration and exploitation of outer space and celestial bodies, centering on spacecraft and space launch vehicle design, manufacture, experiment, launch, operation, return, control, management, and utility. Sometimes astronautical engineering also means certain large-scale space activities, research

Intelligent Testing, Control and Decision-Making for Space Launch, First Edition. Yi Chai and Shangfu Li.
© 2015 National Defense Industry Press. Published 2015 by John Wiley & Sons Singapore Pte Ltd.

tasks, or construction projects. Usually, theories and methods of systems engineering are adopted to guarantee the implementation and progress of astronautical engineering. One of the most fundamental and essential factors in astronautical engineering is testing, launching, and control for spacecrafts and space launch vehicles.

Also called space vehicle or aircraft, spacecraft can be generalized as all kinds of flying machines that orbit in the space according to celestial mechanics under the particular tasks of space exploration, exploitation, and celestial body research. The first spacecraft in the world is Sputnik 1 launched on October 4, 1957, in the Soviet Union.

Space launch vehicle, also known as space carrier rocket, is a general term for carriers that carry payloads from the ground to a specific location in the outer space or the other way around and from one spot to another in space. Space launch vehicles include disposable carrier rockets and partly-recycled and fully-recycled launch vehicles.

1.1.1 Overview of International Carrier Rockets and Spacecrafts

No power could enable objects to realize cosmic velocity until modern rocket technology emerged. Human beings were being fascinated by space travel in the early ancient time. Thanks to celestial mechanics, scientists were able to study the movement of celestial bodies in the perspective of dynamics and help space pioneers to overcome the gravity so as to lay a theoretical foundation to space travel. In the late nineteenth century and the early twentieth century, Konstantin Eduardovich Tsiolkovsky, a Russian and Soviet rocket scientist and pioneer of the astronautics, theoretically proved it possible to overcome gravity to go into space by using multistage rockets. He put forward three important concepts—the first cosmic velocity that is needed to orbit around the earth; the second cosmic velocity that is needed to break free from the gravitational attraction of the earth; and the third cosmic velocity that is needed to break free from the gravitational attraction of the solar system. He gave the main equations for the kinematic and kinetic equations of rocket. All of his efforts and contributions laid a solid foundation to astronautics and helped space travel come into reality.

After World War II, the United States and the former Soviet Union succeeded to develop medium-range and long-range surface-to-surface ballistic missiles. In August and December 1957, the two countries respectively achieved to launch an intercontinental missile. On October 4th of the same year, the former Soviet Union transformed such missile into a carrier rocket for Satellite 1, the first artificial Earth satellite in the world with which human beings opened a brand new chapter for space research. One month later, using this kind of carrier rocket again, the former Soviet Union launched Satellite 2 into space carrying a little dog named Laika. On January 3, 1958, the United States launched Explorer 1, the first artificial satellite carried by the carrier rocket Jupiter-C which had been designed and upgraded by Wernher Von Braun. In April 1961, the Soviet Union successfully launched Vostok 1, the world's first manned spacecraft using the Vostok carrier rocket. Yuri Gagarin, a Soviet pilot and cosmonaut, became the first human to journey into outer space. In July 1969, the United States launched Apollo 11 by the carrier rocket Saturn V. Neil Armstrong, one of the most famous American astronauts, became the first person to walk on the moon. After the former Soviet Union and the United states were France, Japan, China, the United Kingdom, Europe Space Agency, and India that achieved to launch their own first artificial satellite one after another using carrier rockets they developed by themselves.

With the development of space missions and the invention of new spacecrafts, carrier rocket technology has made a rapid progress. These carriers are called "high ladder" for human beings to get to outer space.

Through the developmental history, it was ballistic missiles that pushed carrier rockets to improvement because the removed missile warhead and some adaptive modification brought carrier rockets into reality.

1.1.2 Overview of Chinese Carrier Rockets and Spacecrafts

On October 8, 1956, China founded the first missile research institute whose president was Hsue-Shen Tsien, the renowned Chinese missile scientist who was assigned to be responsible for missile and space programs of China. From then on, China's missile and space technology started to boom in a rapid way.

1.1.2.1 Development of Missile

China's research and manufacture of missile started in the late 1950s. In June 1964, China achieved success in the launch test for the first ballistic missile, which played a significant part in missile and space research. Then China succeeded in the research and manufacture of medium-range surface-to-surface liquid-propellant missiles, intercontinental missiles, and submarine-launched solid-propellant missiles.

In addition, China made a tremendous progress in research and manufacture of surface-to-air missiles, air-to-air missiles, antiship missiles, and modern cruise missiles. At present, China has developed various types of short-range missiles, medium-range missiles, intercontinental missiles, strategic missiles, and tactical missiles.

1.1.2.2 Development of Carrier Rockets

Since the middle 1960s, China's development of carrier rockets was on the basis of missile programs. Trough arduous research and exploration, China developed Chang Zheng (CZ) series of basic and improved carrier rockets including CZ-1, CZ-2, CZ-3, and CZ-4. It is worth mentioning that the successful launch of CZ-3 marked China as the third country just after the United States and the former Soviet Union to master rocket thrusters of low temperature and high energy from liquid hydrogen and liquid oxygen and as the second country after the United States to master the engine second set-up technology. At present, China's Chang Zheng series of carrier rockets are able to launch various kinds of satellites into low earth orbit (LEO), sun-synchronous orbit, and geostationary transfer orbit (GTO). LEO carrying capacity ranges from 0.75 to 9.2 t and GTO carrying capacity from 1.5 to 5 t. Because of excellent technical inheritance, advancement, and low price, Chang Zheng series of carrier rockets have already found their place in the international space launch commercial market. By June 30, 2007, Chang Zheng series of carrier rockets had completed 100 launches, sending 112 domestic and foreign spacecrafts in orbit. Up to the end of December in 2011, 155 launches had been completed.

1.1.2.3 Development of Spacecrafts

On April 24, 1970, China successfully launched the first satellite DongFangHong-1 carried by CZ-1, becoming the world's third country to independently research, manufacture, and launch artificial satellites.

In November 1975, a recoverable satellite carried by CZ-2 was launched in success, making China the world's third country just after the United States and the Soviet Union to master satellite recovery technology. This exerted positive influence on starting manned space activities.

In April 1984, China launched DongFangHong-2, the first GTO experimental communication satellite carried by CZ-3. The satellite was positioned at 125° east longitude above the equator.

China's meteorological satellites are all named as Fengyun (FY). In September 1988, China launched FY-1, the first sun-synchronous orbit satellite, and then four polar-orbit meteorological satellites and three geostationary meteorological satellites that made great contributions to weather forecast. China is now one of the countries that possess both polar-orbit and geostationary meteorological satellites.

China's spaceships are all named as Shenzhou (SZ). China successfully launched four unmanned spaceships (SZ-1, SZ-2, SZ-3, and SZ-4) respectively in November 1999, January 2001, March 2002, and December 2002. All the four spaceships returned to earth smoothly after orbiting the earth. On October 15, 2003, China's first manned spaceship Shenzhou-5 was launched in success. Liwei Yang became the first Chinese astronaut who traveled into space. With such accomplishments, China was marked as the third country after the United States and the former Soviet Union being able to independently conduct manned space activities. In October 2005, China launched the second manned spaceship Shenzhou-6, which realized the first multi-manned and multi-day spaceflight. During September 25 and 28, 2008, the manned spaceship Shenzhou-7 fulfilled its flight task and the first Chinese extra-vehicular activity was realized during this task. The unmanned spaceship Shenzhou-8, launched on November 1, 2011, was automatically docked with the Tiangong-1 space module (launched on September 29, 2011) 2 days after its lift-off. This unmanned docking had orbited 12 days before Shenzhou-8 separated from Tiangong-1 and then docked with it again. This proved that China was able to master core technology of space rendezvous and docking and docking flight.

Tiangong-1, composed of an experiment module and a resource module, is China's first target aircraft with a total length of 10.4 m and maximum diameter of 3.35 m. Its successful launch meant that China started to construct the first space laboratory. On November 3, 2011, Tiangong-1 was automatically docked with Shenzhou-8. Then on June 16, 2012, the manned spaceship Shenzhou-9 was launched triumphantly. Two days later, it was automatically docked with Tiangong-1 and then was manually docked on June 24.

China's global navigation satellite system is named Beidou. From 2000 China began to build Beidou navigation experimental satellite system (Beidou-1) and Beidou navigation and positioning satellite system (Beidou-2). The first and second satellites of Beidou-1, launched respectively on October 31 and December 21, 2000, made up integrated dual satellite navigation and positioning system that provided all-weather, all-time satellite navigation and positioning information. Beidou navigation and positioning satellite system (Beidou-2) was started to be built in 2007. Its first satellite was successfully launched on April 14, 2007. The 10th Beidou satellite was launched on December 2, 2011. Beidou system is regarded as the third complicated global navigation and positioning satellite system besides American GPS and Russian GLONASS.

Chinese people finally realized to fly to the moon when Chang'e-1, a lunar probe was successfully launched on October 24, 2007. On the day of October 1, 2010, Chang'e-2 was launched into space as planned, sending back clear moon surface and polar region image data. It is no doubt that China has experienced significant development in lunar exploration program.

Above all, China made enormous achievements in space exploration and research from the 1950s to the beginning of the twenty-first century.

1.2 Testing and Control System for Space Launch

The role of testing and control system for space launch is to optimize and control the launch process by collecting key data and analyzing information of testing, TT&C, meteorology, and service support. Being technically difficult and systematically complicated, space launch covers numerous scientific domains. As a large-scale comprehensive system with high risks, it includes all essential ground facilities and equipment for carrier rockets, spacecrafts, and their launch together with testing scheme and procedure. It also includes all facilities and equipment required for testing, command, diagnosis, and routine maintenance of carrier rockets and satellites. Typically, testing and control system for space launch is a large-scale and complicated multi-person, multi-machine, and multi-environment system.

The testing and control system for space launch was originated from the system for missile launch. In World War II, Germany took the first step toward rocket use in battle fields. It manufactured the famous rocket V2 in order to make a remote attack on London. In spite of the low guidance accuracy, V2 still manifested its potential destructive power. Testing and control technology for missile launch in the early stage was very simple, so strictly speaking, such technology could not be seen as an independent testing and control system. At that time, launch testing was mainly dependent on simple manual work. After the lift-off, the missile could only make its flight and attack as planned dependent on its auto-control system. Engineers and scientists were not able to track and test the flying missile yet.

After the 1950s, the missile testing and launch technologies were considered more important and great efforts were took to increase launch reliability and rapid response ability. Afterward, the testing and control system for missile launch evolved from manual testing to computer automatic testing, becoming a more complicated system composed of testing subsystem, measurement subsystem, early warning subsystem, information transmission subsystem, and command subsystem.

The main function of testing subsystem for carrier rocket launch is to make tests of all other subsystems before launch and guarantee normal operation of all other subsystems. Its second function is to provide command staff with state information of carrier rockets for decision-making.

The main function of measurement subsystem for carrier rocket launch is to track and measure the carrier rocket in flight, and get flight data like speed, distance, height, and orientation by radar and optical equipment to help describe the rocket flight trajectory. Another function is to, by telemetry equipment, receive rocket state parameter usually including accelerometer value of inertial device, angle data of inertial navigation system, attitude signal, feature parameter, engine nozzle pressure, temperature data, and so on. All the received information helps researchers to know and judge operating state of the carrier rocket in flight.

1.2.1   Components of the Testing and Control System

At present, the world's testing and control systems for space launch are usually composed of various subsystems with different functions. As shown in Figure 1.1, these subsystems are launch testing, TT&C, communication, meteorology, and service support. They are responsible for organizing and commanding, unloading carrier rockets and spacecrafts (satellites, explorers, etc.) from a vehicle, hoisting and transshipping rockets, testing satellites, fueling and launching, measuring and controlling carrier rockets in flight, and providing communication, meteorological, and service support for launch missions.

1.2.1.1   Testing and Launch Subsystem

The primary task of this subsystem is to fulfill the mission of satellite and carrier rocket testing and launching that is completed in the technology area and the launch area. Some tasks are completed in technological area such as unit testing for satellites and carrier rockets, attitude control propellant filling, and satellite propellant filling. Some tasks are completed in launch area such as hoisting, docking, testing, fueling, and launching missions after satellite and carrier rocket transition.

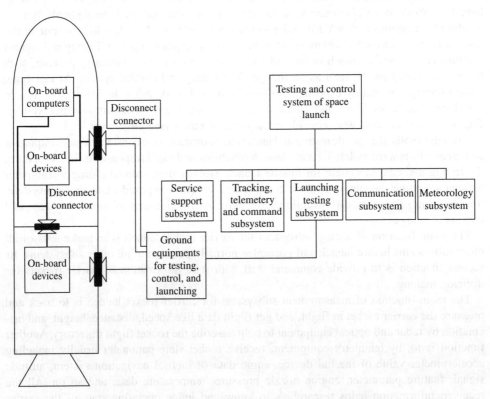

Figure 1.1   Testing and control system of space launch.

1.2.1.2 Tracking, Telemetery and Command Subsystem

This subsystem is designed to complete exterior trajectory measurement and rocket take-off drift measurement. Besides, the subsystem is responsible to receive, record, transmit, and process telemetry data, and to monitor in real time and judge flight conditions of carrier rockets. If fault of emergency happens, safety control measures should be carried out. To sum up, the primary task of this subsystem is to provide command center with monitoring and measuring information.

1.2.1.3 Communication Subsystem

The priority of this subsystem is to provide all the other subsystems with command information, test information transmission channel, and provide unified time service for launch. Unified time service means to provide standard frequencies and time signals for all systems involved so that all subsystems can operate under the unified time benchmark.

1.2.1.4 Meteorology Subsystem

This subsystem has two core tasks. The first one is to provide medium-short term weather forecast for satellite or carrier rocket transition and launching which is important reference for launching plans. The second is to forecast disastrous weather like lightning storms and monitor meteorological conditions in real time for launching before and when some important and dangerous operations are in process like explosive initiating device testing, propellant loading, and launching.

1.2.1.5 Service Support Subsystem

This subsystem offers necessary service and support for launching missions, including railway, road, and air transportation and the supply of barracks, fuel oil, gas, materials, water, electric power, fire protection, medical care, etc.

1.2.2 Functionality of the Testing and Control System

The three functions of testing and control system for space launch are as follows:

1. The first is to test subsystems of launch vehicle and spacecrafts and collect related testing data for analyzing subsystem operation and coordination and for judging whether tested systems meet the design requirements.
2. The second is to organize and command spacecraft launch—decide the feasibility of spacecraft launch according to the operating condition of all subsystems.
3. The third is to monitor and control the flight of launch vehicle and spacecrafts in real time. The system can receive real-time parameter of all subsystems' operating condition by optical device, radar, and telemetry equipment to decide whether the flight is going well. If it is necessary, it can take safety control over carrier rockets- or orbital maneuvers of satellites.

The high accuracy, reliability, and rapidity of space launch make the complexity reach above the rank of 10^6. With launch missions becoming more diverse, large-scale, and complicated, types of data have experienced massive growth. Furthermore, the testing and control system for space launch has its own particularity that other complex systems do not have, such as irreversibility, high risks, disposability of parts and components, and the close connection between the system and the natural environment. The particularity, as well as the uncertainty in testing, launching, and controlling, requires that the testing and control system for space launch should get higher performance.

Researchers need to apply all advanced control theories and methods, such as signal detection and estimation theory, multisource information fusion theory, complex system theory and method, advanced control theory, fault diagnosis, and artificial intelligence in order to analyze key factors that impact safety, reliability, and robustness of system operation. Researchers and engineers are liable to establish automatic testing and control system for space launch through intelligent information collecting, processing, analyzing, and optimal decision-making.

Four main stages are involved in testing and control for space launch.

1. *The Stage of Testing* In this stage, different tests are carried out for launch vehicle and spacecrafts, such as function testing, unit testing, subsystem testing, integrated testing, and functional check. The aim of these tests is to estimate each function index of launch vehicle and spacecrafts and to decide whether launch requirements are met.
2. *The Stage of Fueling* In this stage, staff in charge staff should finish all the preparatory work before the launch, including propellant checking and propellant filling. Dependent on testing data, staff in charge can estimate each function index of launch vehicle and decide whether to continue the launch or not.
3. *The Stage of Satellite and Rocket Flight* This stage is also called active flight stage of carrier rockets. The aim during this stage is to assure personnel and property security and guarantee flight safety during the active flight of launch vehicle.
4. *The Stage of Orbiting* During and after orbit injection, staff in charge should control orbit maneuver and perform given control tasks according to the spacecraft's mission.

This book will talk about only the first three stages and discuss new theories and technology of intelligence systems. The success rate of space launch must be raised substantially if these theories and technology can be applied in the testing and control system.

1.2.3 Technological Processes

Technological processes are scheduled in a reasonable way according to different stages of testing and control for space launch. The processes include testing technological state of products in each stage, ground facilities and equipment, and scientifically assigning work content of different stages in priority order based on testing methods, testing content, and testing procedures. Technological processes are largely influenced by the following factors: the nature of launch missions, launch objects and their features, models of technological process testing, and modes of launch testing and control.

To assure correct operation, excellent performance, and precise technological state prior to on-time launch, staff in charge should work on comprehensive testing and examine the coordination of spacecrafts, carrier rockets, launch site systems, and ground communication system. Testing and control processes should follow the following principles: device testing, subsystem testing, matching testing, overall checking, and connector checking preceding functional testing. Testing and control processes have the following five stages: (i) unit testing in the technology area, (ii) folding and transition in the technology area, (iii) subsystem matching (subsystem and matching checking), (iv) overall checking, and (v) fueling and launching. Technological processes are shown in Figure 1.2.

Therefore, technological processes of testing and control for space launch are very important to success rate and launch efficiency. The processes determine the technological procedures, key technology state, and major work flow of spacecraft and carrier rocket launch; show the relations and work sequence among all subsystems and independent projects; display the time arrangement, key nodes of quality and safety control, and joint operation of the whole system; and finally provide safety measures, the last work tasks, and launch procedures in the launch area.

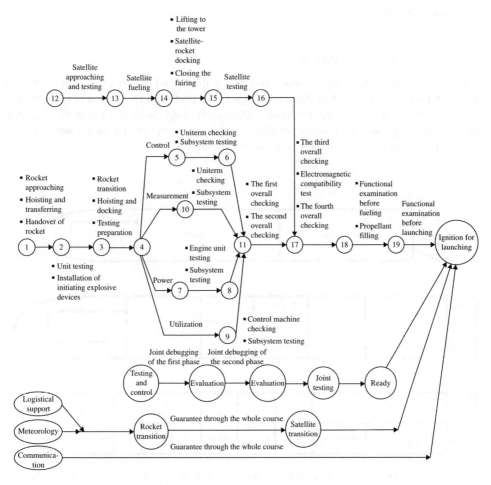

Figure 1.2   The testing, control, and launch process.

Apparently, technological processes are the most important overall technological scheme that organizes and commands the launch of spacecraft and its launch vehicle. The scheme stipulates methods of checking, assembling, testing, docking, task transition, and coordination. It also shows launch objects' technological and launch preparatory work, the work flow and joint operation of launch implementation, and safety and reliability measures, which links closely to launch nature, launch objects, and the development of automatic control technology, communication technology, electronic technology, and computer technology.

There are many factors that influence the processes, including the following ones.

1. Testing objects should be determined reasonably and scientifically; test coverage requirements should be met; and five principles of quality problem close loop must be followed.
2. To meet launch requirements, spacecrafts and launch vehicle should be tested and checked at the system level. System tests include leak detection, precision retests, function retests, resistance retests, mass property tests, and polarity checking.
3. Corresponding testing state, testing modes and testing rules should be determined in light of testing outline.

1.2.4   Developmental Stages

Testing and control technology for space launch was developed mainly by the United States, the Soviet Union, the United Kingdom, and France. The study and application of space launch technology were originated from missile launch technology. With the progress of launch vehicle technology and the raise of reliability and security, the carrier rocket structure has changed from simplicity to complexity, as shown Figure 1.3.

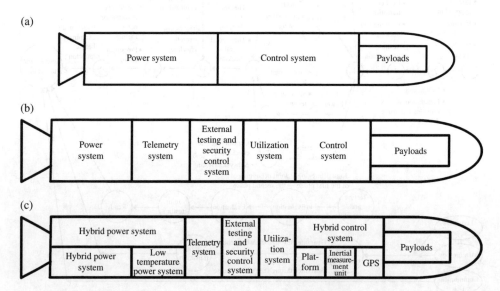

Figure 1.3 Structural evolution of carrier rockets. (a) Simple rockets, (b) relatively complicated rockets, and (c) complicated rockets.

1.2.4.1 Manually Operating Testing

Manually operating testing means that carrier rocket testing and control depends largely on manual work. Technicians make tests of spacecrafts and launch vehicle manually with multimeters, signal generators, and oscilloscopes and then get the readings always by manual methods, as shown in Figure 1.4. The disadvantages of manual testing are low efficiency and high incidence rate of human errors. However, simple system and low price are its advantages.

1.2.4.2 Electromechanical and Analogous Automatic Testing System for Missile Launch

The United States, Soviet Union and Europe exerted great efforts on research and develop fast testing and control system for missile launch. Engineers designed the specialized testing console by which they could test spacecrafts and launch vehicle and simulate computer automatic testing and control system according to specialized electromechanical testing programs.

1.2.4.3 Automatic Testing and Control System

Computer automatic testing and control technology enjoyed a rapid growth due to computer technology development, low price of computer products, excellent computer performance, and myriad practical advanced computer language. Such automatic systems were represented by MARTAC automatic missile testing system developed by Lockheed Martin of the United States, Saturn V automatic testing system developed by NASA, Universal Test Equipment (UTE) developed by BAE Systems of the United Kingdom, Automatic Test System Series (ATEC) developed by GIFAS of France, and POCCNET-distributed testing system developed by NASA/GSFC.

1.2.4.4 New Generation of Automatic Testing System

Possessing information sharing and interaction architecture, new generation of automatic testing system is featured by distributed comprehensive testing, online automatic testing, intelligent fault detection, and assistant decision-making. It has realized inter-module, inter-system, system-and-external-environment information sharing, and seamless interaction. At present,

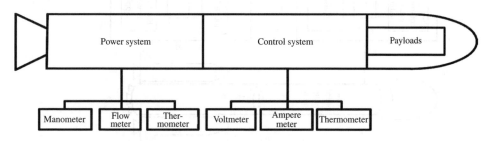

Figure 1.4 Manual testing system of carrier rockets.

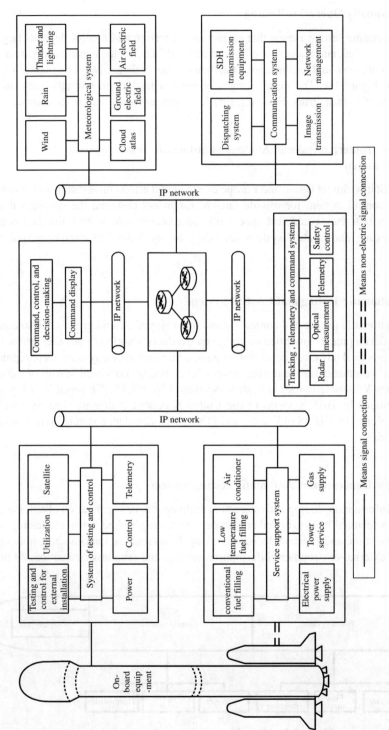

Figure 1.5 System of testing and control for space launch.

abiding by the rule of "Tri-lization" (universalization, serialization, and modularization), the United States uses on-board bus technology (on-board computers do the ground testing of the carrier rocket) to increase testing efficiency and shorten testing period. Both French series of carrier rockets Ariane and American Titanrocket have had telemetry system as their important data-colleting method. They can receive wireless testing data and monitor initiator system, power system, control system, and environment testing system. To raise testing data's reliability and avoid the environmental interference when telemetry information is transmitted by electromagnetic wave, scientists added a project that PCM data stream is transmitted to ground-receiving system through coaxial cables. China has initially realized "Tri-lization" on the carrier rockets CZ-2C, CZ-2D, CZ-4A, and CZ-4B.

After the twenty-first century, there emerged some new features of testing and control system for space launch, like "three verticals" (vertical assembly, vertical testing, and vertical transportation) and the transformation from close control to remote control. The application of computer automatic testing system dramatically raised the testing efficiency, speed, precision of data processing as well as testing reliability. With such new technology, human beings are more and more obsessed with space exploration. In the future, rocket technology will be more advanced and complex; the system will be more complicated; the reliability of carrier rockets will be promoted in a dramatic way. The carrier rocket will gradually become a comprehensive system composed of many subsystems including testing and control, power supply, telemetry, and safety control. In addition, the control system will mature with on-board computers, inertial device, accelerometers, program distributors, servomechanism, and power amplifiers. The control and command system will also develop from electronization to advanced C^4I network integration, leading to the progress of large-scale testing and control system for space launch shown as Figure 1.5.

1.3 Application of Intelligent Techniques for Space Launch Testing and Control

In the twenty-first century, human beings have had full access to space and its exploration. But people need to find more cost-efficient, safer, and faster ways to explore and utilize space resources. Apart from high performance, high reliability, high frequency and high precision, space launch still has many other features shown as follows:

1. Space launch is an industry with high risks, which means any facility or operational fault might result in launch failure, huge national losses, and enormous casualties.
2. Timeliness is one of the most obvious features of space launch. Every space launch is conditioned by the launch window which is usually shorter than 2 hours and "zero window" for the shortest. The launch has to be postponed or canceled if there are serious problems with launch facilities and the window time is missed. Every time after the cryogenic propellant filling, the rocket must be launched within the scheduled time. If the launch needs to be rearranged because of fuel leaking, perhaps it will be postponed for several days or even be canceled.
3. Space launch has many unusual features that other complicated systems do not have, such as irreversibility, disposability of parts and components, strong physical constraints, and the close connection between the system and the natural environment.

4. The equipment structure has become increasingly complicated and automatic with more and more components and parts whose reliability determines that of the whole system and equipment. If a single part or component goes wrong, a heavy loss will be caused. So the number of variables related to system safety and reliability is largely increasing, which poses challenges to safe space launch.
5. In spite of existing mathematic tools, system modeling is still not easy from the perspective of system mechanism because of complicated internal mechanism and numerous influential factors. The whole system is not only featured by nonlinearity, but the coupling relationship among its subsystems is also complex and sometime changes with time.

At present, the testing and control system for space launch has changed from computer automatic testing to intelligent testing. Centering on testing, control, and decision-making for space launch, engineers, and researchers are trying to combine control theories with intelligent methods in order to study advanced control theories and methods like the theory of signal detection and estimation, the theory of multi-sensor information fusion, and artificial intelligence in an interdisciplinary way. To realize intelligent control and decision-making for space launch, engineers and researchers should try the best to master characteristic analysis, modeling, and control decision-making of complicated objects during the launch process and in the control system, helping to establish the theoretical system of modeling, control, and decision-making for space launch.

1.3.1   Intelligent Testing, Control, and Decision-Making Systems of Space Launch

The development of artificial intelligence and intelligent data processing as well as the combination of computer and control technology has brought great progress in the aspects of automatic measuring, intelligent processing of measuring results, and computer control. As mentioned previously, some important parameters of certain controlled objects need to be analyzed, which helps to get a better understanding of the operation and performance of all components and subsystems. With accurate relevant information, controlled objects can be better known in an all around way. If testing analysis and processing is carried through during a physical process, it is called online testing; if beyond or after the process, it is called offline testing.

People always have different understandings about the conception of "intelligence." Usually, people define "intelligence" as the ability changed with external conditions to solve problems and determine the correct behavior by using existing knowledge. In another words, intelligence refers to the correct analyzing and decision-making ability changed with external conditions. Intelligence tends to be reflected by observation, memorization, imagination, contemplation, and judgments. Reasoning, learning, and association are the three elements of intelligence. Reasoning is a form of thinking that can logically make a new judgment (conclusion) based on one or more existing judgments (premises). Reasoning contains two ways: inductive reasoning (from the specific to the general) and deductive reasoning (from the general to the specific). Learning refers to the consistent change of the knowledge structure with environment changes. There are four ways of learning: rote learning, guided learning, case learning, and analogical learning. Association is to know objective things and solve practical problems actively through knowledge combination.

On one hand, besides classical theories and methods of testing, processing, and controlling, the technology of artificial intelligence is also applied to intelligent testing and information processing. On the other hand, there exists manual intervention (such intervention reflects human perception and experience about the environment) through the human–computer interaction in the process of testing, information processing, and controlling. Both of the above two ways can be regarded as intelligent testing and control. Due to the large scale, complicated constraints, multitasks, and uncertainty of the space launch system, the computer should be utilized and explored to the full to achieve artificial intelligence in the aspects of measuring, processing, performance testing, fault diagnosis, and decision output. Artificial intelligence can simulate or perform functions related to human thoughts and be applied to intelligent testing, control, and automatic diagnosis. Artificial intelligence is expected to get the most satisfactory results with the least human intervention. Being featured by fast testing speed, strong processing ability, reliability, and convenience, artificial intelligence can be conductive to the integration of test processing, diagnosis, and decision-making.

Intelligent testing and control technology for space launch can help people to obtain testing information automatically by using relevant knowledge and tactics. Appling technologies of real-time dynamic modeling, controlling, artificial intelligence, and decision-making can be beneficial to testing, monitoring, control, and diagnosis of space launch. The safety and performance of tested objects will be effectively enhanced if intelligent testing and control technology is applied. Meanwhile, the launch system can have high reliability, maintainability, anti-interference capability, adaptive capacity as well as excellent generality and expansibility. In conclusion, featured by intelligentization, based on control theories, and centering on testing, diagnosis, control and decision-making, the intelligent testing and control system is a comprehensive technology that applies and combines the theory of signal detection and estimation, the theory of multi-sensor information fusion, advanced control theory, computer technology, data communication technology, reliability engineering, artificial intelligence and expert systems, intelligent decision-making, etc.

As shown in Figure 1.6, the intelligent testing and control system is made up of six sections: parameter testing, network transmission, comprehensive data processing and analyzing, launch control, fault diagnosis, and situation assessment and intelligent decision-making.

1.3.1.1   Parameter Testing

The section of parameter testing is made up of testing and measuring equipment. There are three types of testing signals.

The first one is analog quantity. All testing analog signals are converted into electric signals by related sensors. Certain analog signals are converted by the A/D converter into digital signals that computers can receive and process easily. The converted digital signals are transmitted to computers by testing equipment through networks.

The second one is digital quantity. Some digital quantities that need to be tested are converted into binary signals by sensors. To be adaptive to the interface circuit, such converted signals need to be amplified or decreased and then are transmitted to computers by testing equipment through networks.

The third one is switch quantity. Switch quantity means the abrupt voltage signal generated at the moment when the travel switch is on. All switching signals that need to

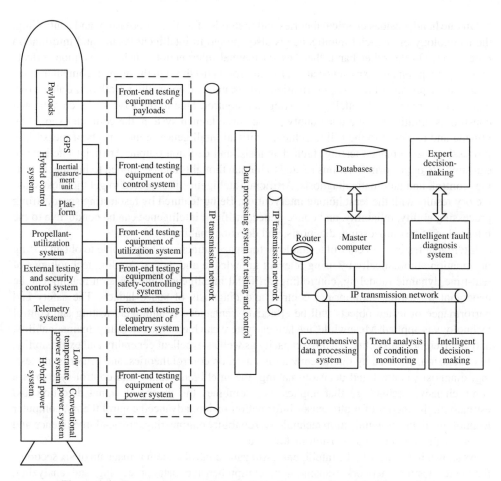

Figure 1.6 System structure of intelligent testing, control, and decision-making.

be tested are converted into DC voltage by testing equipment and then are transmitted to computers through networks.

1.3.1.2 Network Transmission

Considering the features of the on-board control system and relevant ground testing systems, data bus architecture (such as bus network 1553B) is supposed to be adopted because of the advantages of high reliability, good real-time control, and excellent compatibility. Hence, it is easier to achieve bus-oriented and integrated design of ground testing, launch control, and comprehensive diagnosis. The technology can effectively simplify the system, reduce procedures to improve reliability and efficiency, shorten the launch period, increase launch success rate, and promote the system's overall performance by helping to realize miniaturization, modularization, generalization, intelligence, automation, and visualization of the integrated testing system.

1.3.1.3 Comprehensive Data Processing and Analyzing

The integrity, uniformity, and correlation of parameters need to be analyzed on the basis of testing data and telemetry data. Also, the quality of data processing and analyzing can be improved in light of multivariate data fusion.

1.3.1.4 Launch Control

Relevant testing and preparatory work should be done before the launch, including giving manual or automatic backend orders, offering frontend responses to the back end, and controlling the performance of on-board equipment.

1.3.1.5 Fault Diagnosis

Fault diagnosis for carrier rockets has two aspects. On one hand, fault diagnosis, trouble-shooting, and performance assessment for the carrier rocket depend on testing data from sub-system testing and integrated testing with which the possible faults can be tested, diagnosed, and located in time. On the other hand, analyzing parameter variation and working state of the carrier rocket depends on telemetry data in order to support the analysis and decision-making of safety control.

1.3.1.6 Situation Assessment and Intelligent Decision-Making

Analogue simulation and situation analysis should be carried out for the launch and flight safety. An integrated rescue system in the launch site should be established, and a fault handling scheme as well as emergency measures needs to be worked out to increase the ability of emergency security and real-time decision-making.

1.3.2 Key Techniques for Intelligent Testing, Launching, and Control

1.3.2.1 Comprehensive Data Analysis and Intelligent Information Processing

Time and frequency domain data processing methods are taken to test, select, and process all types of testing data and to accomplish normal data extraction and abnormal data removal by weeding out singular terms, trend terms, periodic interference, and noise interference without the influence of zero drift caused by sensors and convertors. Such integration and classification can solve the issues of data consistency and integrity, and can offer effective data support for data fusion, intelligent interpretation, the analysis of reliability, and assistant decision-making.

It is important to study feature extraction methods and algorithm of various parameters and obtain integrated and accurate information about system parameters to build a performance assessment model for data measurement on the basis of correlation models of hierarchical parameters that contain component parameters, subsystem parameters, and system parameters. Moreover, real-time online learning of newly added samples can further perfect the quality assessment model so as to achieve intelligent assessment of measured data.

1.3.2.2 Intelligent Diagnosis Methods and Comprehensive Performance and Quality Analysis

Studying small samples that are suitable for space launch, analyzing features of strong interference data samples and discovering feature extraction methods, researching dynamic feature extraction of the state data and feature reconstruction of the state mode, and establishing a correlation model according to fault concurrency and correlation can offer reliable data and information for the safe operation of the system, fault diagnosis and analysis, and reliability analysis. In addition, predicting the changing trend of relevant on-board components, units, and subsystems, diagnosing potential faults in all aspects of the whole system, and analyzing the feasibility of equipment reuse can provide scientific guidance for system maintenance and transformation.

Based on the cross-coupling and constraint relations among subsystems and on hierarchical relations among all levels of the system, a model for monitoring and evaluating the system's general performance must be established, and a fast algorithm for abnormal feature classification of the system data must be presented in order to accomplish online assessment of system performance and component operating condition.

1.3.2.3 Multisource Information Fusion of Launch Testing and Telemetry

Under the guidance of the distributed and heterogeneous multisource information fusion theory, with the characteristics of high diversity of testing data and inconsistency of data resolution, a unified description method for multisource and heterogeneous information needs to be presented on the basis of feature space description and feature extraction of multisource and heterogeneous information. In addition, temporal and space alignment of measured data and unstructured information complementarity and integration can improve the accuracy and reliability of information fusion under the conditions of time varying, complexity and uncertainty.

During the space launch and flight, telemetry and outer parameters can show the flight status and the flight trajectory. Changing trends and prediction methods of key parameters from the launch to the second-stage separation should be carefully studied because accidental errors of any single information source will have an enormous impact on data analysis. Fusing effective information from multi-parameter sources can help to avoid accidental system errors effectively, reduce uncertainty, improve the system's detection performance, and increase the reliability and fault-tolerance.

1.3.2.4 Intelligent Control and Decision-Making

Considering the complexity, irreversibility, and multi objectives of the space launch system, researchers and engineers should further study the issues of characteristic analysis of complex objects, modeling, and control decision-making. They can know the performance of the system under test and control and work out the distributed decision-making method that will work in the testing and control system by analyzing the dynamic data and information features at all levels. They need to find out the method of centralized goal decomposition, the handling method of coupling constraints, and the distributed decision-making method. All of the efforts are made for the development of intelligent testing, control, and decision-making for space launch.

2

Networks of Testing and Control for Space Launch

The networks of testing and control for space launch refer to the testing and control network and the data transmission network for the launch of space crafts and launch vehicle. Information sharing and seamless interaction are covered in all parts of a single testing system, among different testing systems, and between the testing system and its external environment such that intelligent diagnosis and decision-making can be possible. In the 1950s, the testing and control system for missile launch was built initially, and then were the manual testing system, the semi-automatic electromechanical analogy testing system, and the automatic launch testing and control system. Because of enlargement and complexity of spacecrafts and launch vehicle, testing and control technologies should tend to be networked, automated, intelligent, and integrated. These qualities can contribute to establishment of network architecture featured by information sharing and interaction. More importantly, fast and accurate intelligent fault diagnosis and overall command can be realized.

This chapter will introduce bus networks for testing, launching and control on the basis of the testing and control system for space launch. Focusing on C^3I systems, the author will give an introduction to overall command networks for testing, launching and control, present the automatic testing network and the remote control network in bus networks, and discuss network reliability design, real-time design, and network performance analysis. Besides, to show readers how to achieve overall command and build transmission interconnection network for command systems, the author will provide practical examples of testing and control networks based on 1553B bus, LXI bus, and PLC technology.

2.1 Overview of Testing and Control for Space Launch

Space launch is a large-scale and complicated system engineering of high integration and high risks, covering enormous scientific fields and various systems and technologies. For example, the launch of carrier rockets requires to effectively combine the key data of carrier rocket

Intelligent Testing, Control and Decision-Making for Space Launch, First Edition. Yi Chai and Shangfu Li.
© 2015 National Defense Industry Press. Published 2015 by John Wiley & Sons Singapore Pte Ltd.

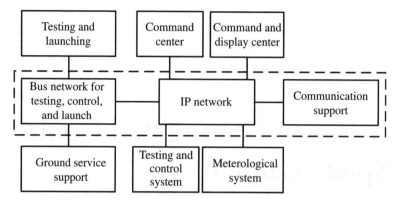

Figure 2.1   Transmission Internet work of testing, control, and command system for space launch.

testing and diagnosis; integrate testing, control, digital display, networks, data bases, fault diagnosis, monitoring systems and power distribution systems, fuse testing, control and data processing of each system to form a distributed processing system through bus networks; and finally realize integration of testing processing, diagnosis, and decision-making. As shown in Figure 1.6, the testing and control system for space launch has two parts: bus networks and overall command networks for testing, launching, and control. Bus networks are featured by real-time testing and control, whereas overall command networks refer to transmission interconnection networks of the command system which contribute to overall decision-making and command with information sharing and interaction among several sections like meteorology, service support, command, and dispatch, as shown in Figure 2.1.

2.1.1   Bus Networks for Testing, Launching, and Control

2.1.1.1   Testing and Control Technology

Testing and control for space launch means that ground facilities control the overall testing and launching process when spacecrafts and launch vehicle are in technical preparation and are about to be launched at the launch site. All the spacecraft and launch vehicle testing, launching, and control strategies fall mainly into two categories. The first one is close-range testing and control and the second one is long-range testing and control.

Close-Distance Testing and Control
This technology was the first to be taken in the field of spacecraft and launch vehicle testing, launching, and control. In the launch site, there are two testing areas, each of which has a set of testing and launching equipment to do horizontal and vertical comprehensive tests and to launch the spacecraft and its carrier. In the launch area there is a solid and secure underground launch control room in case of injuries from sonic boom, thermal radiation, and accidental explosion.

In such technology, a technology area and a launch area are set up in one launch site, similar to the mode of testing and launching process which includes a testing area and a launch area. Thus, spacecrafts and their carriers can be tested horizontally and vertically.

Long-Distance Testing and Control

In the 1960s, the United States and the Soviet Union developed the technology. The United States even used computers in one control center to control the testing and launch of several carrier rockets. Due to the technology, it is not necessary any more to keep the solid underground launch control room. Instead, a testing, launching, and control center is set up in the technology area. Testing and launching equipment in the center undertakes overall testing in the technology area as well as in the launch area. The center aims to take the long-distance control for the launch, combining the technology area and the launch area.

Currently, countries that adopt the three-vertical model (namely vertical assembly, vertical testing, and vertical transportation) generally make use of the long-distance testing and control technology, which is recognized as the main trend of testing and control for space launch.

China's long-distance testing and control technology for space launch is coupled with the three-vertical model in testing and launch process. Such advanced technology will lead to the establishment of highly automatic launch sites. The technology area has been deemed as the key part in the launch site since the three-vertical mode and the long-distance testing and launching technology were adopted. The technology area and the launch area are more closely linked by vertical transfer tracks and information networks. All the work of the testing and launching control, like testing, fueling, fire protection, and monitoring of cable swing beam and the aimed window, are under the control of the testing and launching command system built on computer networks. The center is also responsible for the assistant decision-making if there is any emergency during the ignition, such as fire, propellant leaking, or power supply problems. Fueling, air feed, air conditioning, non-standard device, and waste gas treatment are also automatically controlled by computers, which effectively improve the security and reliability of the testing and launching project.

1. *The Safe Long-Distance Testing, Launching, and Control Center* A testing, launching, and control center with good internal environment can be constructed in the technology area far away from the launching pad, in order to make remote control of the spacecraft and to raise the security and reliability of the launch.
2. *The Advanced Long-Distance Information Transmission Technology* For the spacecraft, information transmission depends mainly on wireless transmission with the assistance of wire transmission that relies more on optical fiber and less on cable. For the carrier rocket, information transmission depends mainly on wire transmission with the assistance of wireless telemetry. The wire transmission also relies more on optical fiber and less on cable.
3. *Long-Distance Control of Launch Service Facilities* In the launch area, facilities of fueling, fire protection, swing rods, and collimation are all controlled and monitored by computer networks which can make long-distance control with high security, reliability, and automation.

2.1.1.2   Testing Technology and Control Networks

Testing technology is developing with the rapid progress of modern science and technology. Space products must be examined and checked with testing technology. No one would deny that testing is an imperative part for product development and manufacture. Space testing technology focuses on the testing and launching of spacecrafts and their carriers, and it includes testing systems and methods composed of hardware and software. The automation degree of testing systems, to a large extent, can reflect overall product performance. In short, testing is an essential means to guarantee the practical performance of space products.

Space Testing Technology

Launch vehicle is a complicated system that is made up of the control system, the power system, the telemetry system, and the orbit measurement system. Launch vehicle's testing and launching particularly embodies main problems in launch vehicle testing and reflects the development trend of testing technology. Studying launch vehicle testing technology, engineers and scientists can not only have a good picture of the basic status and key testing technology but also can popularize the study and application of testing technology.

Space testing refers to the examination and measurement of system parameters and performance on the launch vehicle with the aim of confirming the technology state of the launch vehicle. Usually, bus testing is the first and foremost choice because it can control work parameters and the stability of all system equipment continuously and dynamically. From the perspective of physical distribution, all function blocks that are put in either the front end or the back end can form a centralized-distributed processing structure; from the perspective of software, the multi-layer protocol that allows information sharing is the fundament of testing.

Testing work includes measurements of the power system, output of the control system, and condition parameters of equipment on the launch vehicle. In the next place, the measured information should be converted, collected, stored, transmitted, processed, and displayed in the ground testing, launching, and control system to make sure that functions of power supply and distribution, testing, control, and diagnosis will play in a normal mode. Usually, there are two testing sections—bus testing and on-board testing—and testing parameters of the telemetry system are coordinated well, which means that work parameters and stability of the control system's equipment can be mastered continuously and dynamically. Firstly, there must be a control center that coordinates testing, launching, and control procedures and commands the parallel operation of computers and front-end equipment. Then through integrated data processing, the launch process will be determined in an automatic mode that can be manually intervened in.

During testing data collection, it is too limited to decide if projects are qualified or not just by one single test. All possible approaches (wired and wireless transmission, networks, bus, and historical records) and sources (plant testing data, unit testing data, and historical testing data) should be used to collect all testing data of on-board equipment from the delivery to the launch so as to establish a perfect database. The core part of testing data collection is the diagnosis model for both performance diagnosis and fault diagnosis with artificial intelligence. Based on database, networks, and bus as the system framework, the diagnosis model depends on the cooperation of different function blocks (traditional testing, launching control, and dynamic supply) which are seen as front-end "antenna" for diagnosis. Comparison of previous testing parameters can provide a foundation for judging stability and trend of on-board equipment.

The United States, Russia, and European Space Agency are the three foreign pioneers in the area of launch vehicle testing. There are mainly three transmission approaches—wired transmission, wireless transmission, and wired-and-wireless transmission. Concerning testing technology, the United Sates presented the concept of "Tri-lization" (universalization, serialization, and modularization). In the testing plan, the on-board bus technology has been in full use, and that on-board computers carry out ground testing work has largely raised testing efficiency and has shortened the testing period. European Space Agency has applied wireless transmission technology in testing transmission and has used the telemetry system as a key means of data collection. The carrier rocket "Arianne" just received testing data in a wireless way.

Bus Networks for Space Testing, Launch, and Control

Currently, bus technologies of CAMAC, VXI, LXI, and 1553B are mainly utilized. There exist types of testing systems in different degree of automation due to the structure and equipment diversity of different series and versions of launch vehicle. Comparing testing technology and data transmission approaches of foreign countries, China is supposed to develop testing and network transmission technologies that are adaptive to commercial development of large-scale launch vehicle.

In remote control networks, the mature PLC control technology, field bus technology, and computer network technology have been utilized. Most of the transmission media are shielded cables and single-mode or multimode fibers. Such remote control networks have high reliability, low retardance, and strong anti-jamming capability. With increasing of equipment complexity, increased information transmission, and the requirement of integration of testing, launching and control, it is urgent to present new information transmission plans in remote control networks and build a network architecture that combines measurement, launching control, command, and application.

2.1.2   Overall Command Networks

Constant progress of space testing and control technology as well as improving computer processing performance and increasing IP network transmission speed has led to a great increase in command information and control data and network applications that are based on real-time high-speed data, three-dimensional geographic simulation, and multimedia videos and images. Information transmission of testing, TT&C, and meteorology systems has gradually changed from point-to-point interconnection to IP packet switching. Accordingly, a better information transmission network with enough bandwidth, easy access, and flexible networking is needed to cope with such changes. More importantly, the information transmission network must have the features of high reliability, safety, and real-time effect.

In the area of space testing, launching, control, and command, the study and application of high-speed IP network transmission must be emphasized more if all types of information are expected to be transmitted in a fast and reliable way.

Although the IP network is fast, flexible, cost-effective, and easy to use, it has some serious disadvantages in network security and quality of service (QoS) because of the openness of IP protocols and the service principle of "best effort." If no measures are taken, the IP network will not be reliable, secure, and controllable enough to guarantee the success of space testing, launching, control, and command work. Thus, comprehensive technology and management measures are in desperate need to cover its shortages.

The following are the four issues to be addressed for the establishment of IP network for space testing, launching, control, and command.

2.1.2.1 How to Guarantee High Reliability

There are many factors impacting network reliability, such as the reliability of hardware equipment, telecommunication lines, and network applications like antivirus programs and authentication service. In addition, zip storms, large-scale network virus outbreak, abnormal access terminals, and power supply problems may also result in network anomaly or network shutdowns if any of these problems occurs.

2.1.2.2 How to Guarantee High Speed

Data transmission in space testing, launching, control, and command requires high quality of network service, so time delay or high packet loss rate is hardly allowed. However, IP network uses shared channels instead of dedicated channels. Although such IP network can meet real-time requirements in general, it will not be fast enough when network traffic is heavy.

2.1.2.3 How to Guarantee Network Security

Rapid internet development has contributed to the diversity of network applications based on IP technology and made people's work and study more efficient. At the same time, networks are flooded with virus spread, hacker attacks, fake users, illegal scanning, and information stealing. Such harmful behavior, like Nimaya, is posing increasing threat to normal network order. Task Private IP networks also confront the same network security issues. Some related security measures must be taken to solve these problems.

2.1.2.4 How to Make Network Manageable and Controllable

The launching center's task IP network is built on the basis of Ethernet technology, but Ethernet has its disadvantages in some aspects such as authentication for network access, user behavior management, broadband limit, flow control, and fault warning. It is not able to fulfill delicacy management of task networks.

In the international arena, NASA and ESA have basically realized IP network information transmission, exploring the practical application of IP technology in the field of space testing, launching, and control. The private IP network for space testing, launching, control, and command is an independent private network. Because of user control, business control, and flow control, certain technological measures can be taken to address the issues of reliability and security. Thus, information transmission for space testing, launching, control, and command can be on the rails.

2.2 Bus Network Architecture of Testing, Launching, and Control

The key role of the testing, launching, and control system is to test functions and practical technical index of launch vehicle and ensure the success of ignition and launch. Bus network of the testing, launching, and control system can be segmented into two parts—automatic testing network and launch control network—connecting all subsystems of the ground testing, launching, and control system as well as the on-board fight control system.

Bus network of testing, launching, and control can achieve physical connection among shared testing, launching, control, diagnostic information, and resource. It has three basic structures: master–slave structure managed by a master station, peer-to-peer structure featured by distributed information processing, and client/server and browser/server structure. Because of its particularity, bus network of testing, launching, and control has the following functions: managing the flow of testing, launching and control information among all systems, collecting and sharing data, integrating and simplifying the management and backup of data and software, and realizing distributed processing and integration of all subsystems.

In the following content, the author will introduce bus network architecture of testing, launching, and control in detail by illustrating its network composition and implementation model. Automatic testing, launching, and control network architecture and remote control network architecture for space launch will be taken as practical examples.

2.2.1   Network Architecture for Automatic Testing and Launching

The testing system must be able to fulfill both on-board equipment testing and ground comprehensive testing. There are two issues to be addressed during the network design. The first one is to select between "on-board testing" and "ground testing"; the second is to decide the types of the testing bus.

2.2.1.1   On-Board Testing and Ground Testing

Whether testing modes and bus standards are reasonable has a direct bearing on reliability, complexity, compatibility, and maintainability of the testing and control system. Take the traditional point-to-point connection which uses directly connected cable network to transmit signals between equipment. The complex cable connected network may lead to electromagnetic interference and poor system reliability. Currently, CCAMAC and VXI testing networks are adopted, but they have two disadvantages: huge size and complex cable connection.

Launch vehicle system being increasingly complex, more and more signals need to be transmitted to the ground before they get tested, which will definitely multiply the complexity of connection between the rocket and the ground. Then a series of issues, including realizability, maneuverability, and preparation time for launch, must be paid attention. Above all, related staff should take into consideration the status quo and development history of current testing network architecture before selecting either "on-board testing" or "ground testing" and deciding which bus standard is more suitable.

2.2.1.2   Simultaneous Testing of Rocket-Ground Buses

Rocket-ground simultaneous testing is a system architecture that combines the advantages of both "on-board testing" and "ground testing." When it comes to bus selection, many factors must be taken in account, such as compatibility.

Firstly, it is necessary to select between "on-board testing" and "ground testing." Carrier rockets are autonomous, independent, and disposable products whose self-testing technology should be moderate. Accordingly, "on-board testing" and "ground testing" should both be applied to cover each other's shortages and accomplish the launching and testing task jointly. On one hand, the on-board system can fulfill all the tasks at circuit board level and parts of tasks at equipment level; on the other hand, ground testing can focus on dynamic testing of key signals, transition between testing states, necessary external excitation, and all-round operation of rocket-ground simultaneous testing. Such cooperation and coordination can not only simplify the testing architecture but also improve the testing system's efficiency and reliability.

Secondly, when it comes to bus selection, the principles of maneuverability, extensibility, openness, and reliability must be followed. In addition, a common feasible standard suitable for existing testing equipment needs to be determined to fulfill the integration of various testing equipment, which can guarantee openness and extensibility to the largest extent apart from increasing resource use efficiency and promoting system adaptability.

According to the idea of rocket-ground simultaneous testing, it is possible to build a distributed testing system at three levels in terms of testing subjects: board level (BL), equipment level (EL), and system level (SL). The on-board equipment testing at circuit BL can apply BIT (Built-in Test Equipment) to accomplish self-testing and self-diagnosis. Testing at EL can adopt both BIT and external testing. Testing at SL can follow comprehensive testing structure, which means all ground testing projects should coordinate with each other. The architecture of such on-board and ground simultaneous testing system is shown in Figure 2.2.

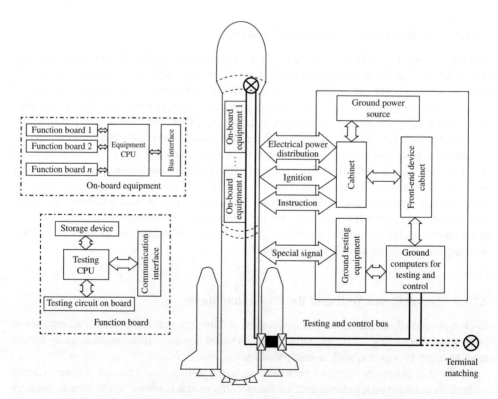

Figure 2.2   Architecture of on-board and ground simultaneous testing system.

Testability design of on-board equipment should be carried out at different levels according to the features of testing subjects. It is also important to balance testability and reliability. There are three levels regarding to testing subjects: BL, EL, and SL.

1. BL products have multiple functions. Apart from digital circuits, BL products also contain analog circuits, mixed circuits, high-frequency circuits, power circuits, and optical circuits. Generally, these circuits may pose obstacles to BIT design. With the help of microprocessors and sampling circuits, BL products can achieve the function of self-testing with certain testing coverage.
2. The purpose of EL testability design is to assemble circuit boards with the self-testing function via the board bus. As long as the bandwidth can support the communication rate, the board bus can be regarded qualified, such as PCI-Express and VXS. When it comes to network topological structure, it has three forms: tree topology, star topology, and bus topology. When the whole system is in operation, auxiliary testing information from each function board is being collected and transmitted to the ground comprehensive testing system and the telemetry system via rocket-ground interface and telemetry interface. Then the data can be comprehensively applied in such way.
3. The purpose of SL testability design is to set up an "information highway" for transmitting control flow information, imperative mutual information, and large amounts of testing data. Most of information channels at the SL are serial buses, for instance, 1553B bus, FC (Field Control) bus, CAN bus, and LXI bus.

Using a single bus in the ground comprehensive testing system is now increasingly inapplicable because of the long-period operation. Therefore, it has been a main trend to develop extensible and open technologies that can reduce the coupling degree among testing equipments to the largest extent. LXI bus and FC bus represent such promising technologies.

Automatic testing and launching testing networks based on 1553B bus and LXI bus enjoy open industry standards, small size, strong interoperability, and high extendibility. Moreover, the LXI technology is highly compatible and suitable to be applied in automatic testing and launching networks since it uses interface modules compatible with GPIB, VXI, and PXI buses. 1553B bus, though no match for the rising LXI bus, is also largely applied in the space and aeronautics industry due to its excellent openness, adaptability, reliability, and fault tolerance. With the optical fiber transmission technology, 1553B bus can be improved to overcome its low transmission rate, which will increase its use in both short-range and long-range testing networks.

2.2.2 Remote Control Network Architecture of Space Launch

Remote control network architecture of space launch has three main parts: front-end testing and control equipment, transmission channels, and the long-range testing and control center. In this section, the author will discuss field bus networks and optical fiber networks in the remote testing and control network that can fulfill long-range monitoring, diagnosis, launch control, and data display.

The structure of the remote control network of space launch is shown in Figure 2.3. The whole structure contains ground testing, launching and control system, master control computers,

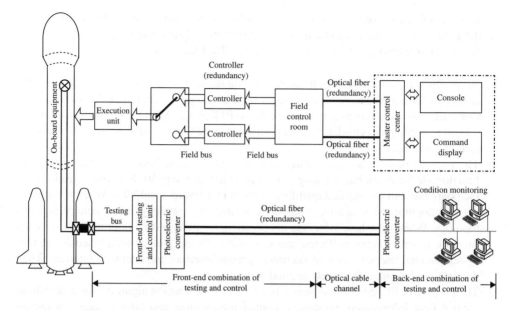

Figure 2.3 Remote control network of space launch.

multimedia management system and its matched data management system. In this structure, field buses are adopted as transmission channels for control information and optical fiber networks for monitoring data. Local testing and launching network architecture, having been introduced in detail in Section 2.2.1, uses VXI bus technology to digitalize remote testing and launching, front-end control, and information measurement, so data is able to be transmitted via communication systems. Front-and-back end data communication must be added to address the issues of control information transmission and measurement information transmission.

In the communication part, field buses are adopted as transmission channels for control information and optical fiber networks for monitoring data. Single-mode optical fibers and single-mode optical transceivers, without any repeater, can transmit signals at the range of 5–7 km, which meets the requirement of long-distance signal transmission.

Focusing on mature and reliable power line communication applied in the industry, the control and launching system has achieved front-and-back end data collection and networked information transmission. In the meantime, information sharing and exchange have also been realized for testing crews to get a clear picture of the testing process and testing information in time.

2.2.2.1 Control Process

The control process is designed to fulfill sequence control based on the launch process including rocket approaching, testing, join-test verification, fuel filling, etc. As a terminal directly connected on the bus, the ground testing and control microcomputer can get a large amount of testing information from on-board equipment, and as rocket-ground data transmission interfaces, they can also check and start on-board computers to get computational results.

2.2.2.2 Controller Redundancy

The function of controller redundancy is to guarantee reliability and success of rocket launch during the window period. While designing a fault-tolerant control system, the designer needs to design dual controllers to create a selective control system, and thus the working controller can be switched over if there is any emergency. It is the master control center that gives control instructions and sends control signals to controllers (redundancy) via optical fibers (redundancy), the field control and transmission system, and the field bus.

2.2.2.3 Long-Distance Transmission

In the remote testing, launching, and control system, the carrier rocket control center and the front-end device room are connected by fiber channels. The current testing and launching technologies have become relatively mature, so these technologies are applicable, only except that sources of testing information need to be changed. Moreover, the previous way ignition exciter signals were sent and the previous power distribution method can also be maintained.

2.2.2.4 The Master Control Center

The function of the master control center is to monitor testing information in real time, analyze abnormal data, and diagnose faults, which means the center has to guarantee normal equipment operation and comprehensively assess the performance of the carrier rocket.

2.2.2.5 Condition Monitoring

The main function of online automatic testing equipment is to test analog data. Mainframe computers for information management have mass storage devices, so they are able to manage a large amount of empirical data, binding data, fault model data, and expert system knowledge bases to which all systems can have access at any moment.

2.3 Key Techniques of Bus Networks for Testing, Launching, and Control

Compared with common bus technologies, on-board testing, launching, and control networks have more distinctive features. It functions comprehensively because of information interaction and its functional subsystems also cooperate and coordinate with each other in an efficient way.

1. Bus networks for testing, launching, and control enjoy higher reliability in hardware and communication protocols. Redundancy, isolation, error detection, and failure recovery mechanism mesh together to lower the bit error rate and prevent network paralysis, which guarantees the normal operation of bus networks.
2. Such bus networks place much emphasis on real-time performance. In these networks, intermodal data transmission should be time-deterministic or time-bound, which means that time limit for data transmission should never be exceeded. The index of bus transmission

delay mainly reflects the real-time performance of a given network. The time of a single transmission can be shortened, thanks to higher transmission rate and shorter information frame.

3. Most of network topological structures are of bus type with simplified connections and higher reliability. It is easier to operate management and error control in such structures. Most communication protocols use the request/response model—the bus controller controls message transmission and related terminals give responses to its command. This model is convenient for central control, thus optimizing bus transmission delay. The buses are required to be featured by high reliability, strong real-time performance, and excellent openness due to the particularity of the testing, launching, and control network. Above all, network reliability and real-time performance should be strictly analyzed and tested.

2.3.1 Reliability

Considering the experience of reliability design and quality control in the design of CZ-2F, one of the Chinese manned carrier rockets, engineers can conduct the reliability design for the testing, launching, and control network based on four aspects: hardware reliability design, software reliability design, reconstruction of network redundancy, and network reliability analysis. In addition, during the reliability design, engineers and designers should also formulate the network reliability, test and analyze the network reliability, and establish a robust system for reliability management.

2.3.1.1 Hardware Reliability

Hardware reliability design refers to providing reliable hardware support at all levels such as BL, EL, and SL. Related hardware support includes derating design, multi-computer backup, interface redundancy, triple modular redundancy, and electromagnetic compatibility design. Current network transmission tends to be distributed and fast, so optical fiber is the best material to be used as transmission medium because of its high bandwidth, high reliability, and good real-time performance, especially when shielded twisted-pair cables have weak anti-jamming capability and composite materials have replaced shells that have shielding function. In short, designers should put a priority on optical fibers as transmission media in the network reliability design.

2.3.1.2 Software Reliability

Software reliability design includes two aspects. The first is to improve the reliability of applications by taking proper measures like information coding, data redundancy, and process control; the second is to carry out fault-tolerant scheduling on different versions of the same application at the level of operating system by using fault-tolerant models, thus keeping the system working in a right way by restarting backup programs if software faults occur. In addition, some design proposals can be seen both as hardware and as software design, which is called collaborative reliability design of hardware and software, such as watchdog circuits, a hardware facility supported by software.

2.3.1.3 Redundancy Reconfiguration for Network Systems

Redundancy rate is a necessary index for a system. Redundancy reconfiguration of network systems means that networks with single-point failures can be granted the ability of automatic restoration and a new communication link can be intelligently rebuilt if three technologies— network switch redundancy, link redundancy, and network interface card (NIC) redundancy— are applied in an integrated way. The technology of redundancy reconfiguration for network systems involves network equipment redundancy, fault detection, and reconfiguration algorithm. Equipment redundancy and reconfiguration design for remote testing and control network systems include the following aspects.

1. For redundancy design inside a single switch, dual power modules need to be applied and ports should be designed to work standby for each other. Moreover, dual NIC model is required inside testing and control equipment.
2. For redundancy design between switches, two front-end switches are required to work standby for each other, and so are the two back-end switches. The six core network links at the front and back ends should also work standby for each other. Dual NIC connected to different network switches via optical fibers or twisted-pair cables should be applied in important testing equipment.
3. The switching system can detect link faults by itself and cut off the fault link automatically before switching to the redundant link, thus accomplishing the reconfiguration of network links.
4. Switches are bound to each other through specific network protocols and work standby for each other. If a certain switch goes wrong, the system can detect it and then cut off the fault or switch to the redundancy.

2.3.1.4 Analysis and Testing of Network Reliability

During the reliability design, "reliability analysis" refers to developing a block figure and mathematical model (the reliability model) for testing products' reliability and failure characteristics through the qualitative method and quantitative method. Reliability analysis plays an important role in the design of the testing, launch, and control network. In the early stage of design, reliability analysis can help related engineers to know exactly the importance of each subsystem and to ascertain righteously the reliability parameter of each subsystem, which exerts a great impact on design optimization and cost saving. In the middle stage of design, reliability analysis can further refine the reliability design of the system and provide quantitative criteria for the design. In the later stage of design finalization, reliability analysis can provide scientific theoretical guidance on whether the system can fulfill the scheduled tasks, so design faults that may lead to mission failure can be avoided. In the network operation, reliability analysis can help to make the quantification of operation efficiency more accurate as well as to find out potential fault sources.

Several factors that may affect network reliability are as follows:

(a) Topological structure
(b) Reliability and maintainability of network parts
(c) Management and control system

(d) Fault-diagnosis capability
(e) Self-recovery capability involving protective methods, repairing methods, and routing algorithm
(f) Operating environment
(g) Users' requirements for network performance, such as throughput capacity and time delay

The concept of "reliability measure" is often mentioned during reliability analysis and testing of the network system. Currently, researches at home or abroad are all based on the idea that a network can be abstracted out as a flow diagram composed of nodes and links through which various types of information are transmitted. Generally, there are four network reliability measures: invulnerability, survivability, availability, and reliability of network parts in a multimode state.

Network Invulnerability
Network invulnerability refers to the minimum number of nodes or links that have to be cut off to interrupt the communication between certain nodes. It reflects the reliability of communication networks when they are being damaged by external environment. As a matter of fact, network invulnerability in question is equivalent to the reliability of the topological structure in communication network. It has two reliability measures: edge connectivity and node connectivity.

Edge connectivity means the minimum number of links that have to be cut off to disconnect all accesses between two nodes, whereas node connectivity means the minimum number of nodes that have to be cut off to disconnect all accesses between two nodes. According to these ideas, edge connectivity can be identified more generally and practically as the minimum number of links or nodes that have to be cut off to get rid of a subnetwork from a communication network; node connectivity can also be identified as the minimum number of links or nodes that have to be cut off to keep network diameter less than or equal to the threshold value.

Presented and formed according to the graph theory, the concept of network invulnerability describes from the angle of network connectivity how topological structure affects reliability of communication networks. Although it cannot describe reliability of communication networks in an all-round way, the concept of network invulnerability describes the reliability of topological structure. It is worth stressing that network invulnerability has nothing to do with the reliability of network parts.

Network Survivability
Network survivability refers to the reliability of communication networks when random destruction occurs. Reflected by probability measure of reliability, network survivability, in a sense, shows the connectivity probability of communication networks. For a communication network, random destruction is often reflected by the survival probability of nodes and links. According to this, researchers have developed many approaches to study connectivity probability of communication networks.

1. End-to-end connectivity probability which has been given the biggest concern refers to the probability of one path at least between two random nodes in network (suppose that nodes in network are absolutely reliable and all links enjoy the same survival probability). Study approaches include state enumeration, the cutset method, and the simple path method.

2. Utilize Monte Carlo method on the basis of survival probability of parts to simulate node and link failures caused by random destruction. Pick out a subnetwork with highest connectivity in the surviving network after destruction, and the percentage of this subnetwork's average node number in the node number of original network is just the connectivity probability of communication networks.
3. In a symmetrical network, utilize Poisson distribution to simulate random destruction in nodes and links and then provide recursive results. Pick out a node randomly in the surviving network after destruction, and the percentage of nodes that can be connected with it in the node number of original network is also called connectivity probability in a symmetrical network.
4. Suppose that nodes in network are absolutely reliable and all links enjoy the same survival probability, and connectivity probability of communication networks can be defined as probability of a connected network composed of the whole communication network. In a broader sense, connectivity probability can also be defined as survival probability of nodes and links in network.

Network Availability

Network availability is a reliability measure concerning network performance such as throughput capacity and time delay. It reflects requirements for network performance when failures of network parts occur.

Scholars have presented several indexes to measure network availability. A simple index of network availability is weighted end-to-end connectivity probability, that is, the average value of the whole network weighted by the corresponding information flow with end-to-end connectivity probability. Another index of network availability is the probability of network throughput capacity exceeding a given threshold value. Such index can well reflect the decrease in throughput capacity caused by failures of network parts. Still another index is based on time delay of network transmission. Suppose that nodes in network are absolutely reliable, that waiting time and processing time of each node are the same and links in network have certain survival probability, that time delay of all links is negligible and time delay between two certain nodes equals to the link number contained in the shortest path between the two nodes. Then network availability can be defined as the probability of transmission time delay from a center node to other nodes not exceeding a given threshold value. In addition, some scholars have also presented availability indexes of circuit-switching networks and message-switching networks.

Reliability of Network Parts in Multimode State

The above three reliability measures concern only two modes of network parts—operation and failure. In fact, performance of network parts will degrade as time rolls on during which there exist many intermediate states. Accordingly, multi-working modes of network parts must be taken into consideration during reliability analysis of communication networks, because this can make results of reliability measure more accurate. However, increasing working modes of network parts and the large amount of network operating states may make computation of network reliability more difficult.

Reliability analysis could be classified into two categories. One is engineering analysis, which means finding the cause of failure through macroscopic and microscopic analysis and knowing propagation regularity of faults before taking corrective and improvement measures.

The other is statistical analysis, which refers to assessing reliability level according to experimental results. Such kind of analysis can ascertain weak links and provide quantitative foundation for design improvement.

Reliability analysis plays a vital role in the whole process of designing, developing, and using products. The technology of reliability analysis can be applied to analyze potential faults of different kinds of products, and then designers and engineers are able to know what measures they may take to guarantee the inherent reliability of products.

It is not hard to see from the state of reliability studies that the reliability of communication networks depends on the reliability of network topological structure, reliability of network parts, and users' requirements for network performance. The most important factor is the reliability of network topological structure, so it is a must to make clear reliability of network topological structure during design. In addition, routing algorithm is another important factor that may affect reliability of communication networks. It must alter the corresponding path dynamically with the change of network topological structure and cope with network congestion in time. Unfortunately, routing algorithm has often been supposed to be ideal and negligible in reliability analysis, which is regarded as a flaw of reliability analysis. Moreover, researches so far has not yet explored deep into communication systems; adopted network models cannot reflect reliability of nodes, and how to introduce routing algorithm into network reliability models remains a problem to be solved. All these problems are the challenges we are facing.

Reliability tests include growth tests and compliance tests. Growth tests at different development stages can reveal weak links, after which pertinent measures need to be taken to enhance reliability and meet users' requirements for network reliability. Compliance tests can verify whether product reliability meets predetermined requirements at different stages. At the end of engineering development, reliability tests are taken exactly as qualification tests.

2.3.2   Real-Time Capability

In the testing, launching, and control network, some semaphores that need to be monitored continuously, and a great amount of control signals require not only reliable network transmission but also high real-time capability. Structurally, the testing, launch, and control network is a distributed real-time system that is required to compute and process within the allotted time and response to outer asynchronous events. In real-time systems, time requirement is the decisive feature for task fulfillment. Real-time systems can fall into two categories according to this feature: hard real-time systems and soft real-time systems. For hard real-time systems, each task has a deadline before which tasks must be fulfilled, or there will be disastrous and irreversible consequences. For soft real-time systems, it is acceptable that tasks are fulfilled after the deadline.

Most of the current distributed real-time systems are hard real-time system. Real-time networks can just meet the strict time requirement in hard real-time systems. It is certain that real-time systems should have real-time operating systems. The most important features of real-time networks are communication certainty and predictability, which means that the time of data transmission between nodes is determinate or, in another word, predictable in real-time networks. Data transmission in real-time networks has time limits. If the deadline is exceeded, this data transmission will be regarded as failure even though the data has been received. A good real-time network should be featured by certainty, predictability, high transmission rate, and strong error correcting capability.

There are two popular categories of real-time networks. One is ring-topological reflective memory networks and the other is star-topological broadcast memory networks. In essence, both networks use high-speed memory mapping and a real-time communication tech-nology to share information and data. These two types of networks have no great difference in working principles and functions but do have some difference in performance index, connection mode, topological structure, communication media, bus support, and operating system.

The following issues should be paid attention to during the design of testing, launching, and control network.

2.3.2.1   Clock Synchronization

Different distributed real-time systems need to have a common time. Each computer has its own local time, so time synchronization is a key issue in design.

2.3.2.2   Event-Triggered and Time-Triggered Systems

Event triggering means that a processing activity is initiated as a consequence of the occur-rence of a significant event, and then a sensor will send an interrupt signal to the connecting CPU. Accordingly, event triggering is also called interrupt driving. For a soft real-time system with light loads, event triggering is a more suitable and effective method. For a relatively complex system, everything can go well if compliers can predict all behavior of the system after the occurrence of an event. The main problem of event-triggered systems is that a large scale of interrupt will be caused if the system load is too heavy.

In a time-triggered system, the activities are initiated periodically at predetermined points in real-time. ΔT refers to the interval time between every two clock interrupts. The interrupt load remains the same, so systems would not be overloaded. It is very important to determine ΔT. If ΔT is too short, too many interrupts in the system will result in a waste of time; if ΔT is too long, emergencies may not be handled in time. Besides, some events may occur in the middle of two interrupts and they must be saved for processing.

In short, event-triggered systems are appropriate for light-loaded work, while time-triggered systems are only appropriate for a changeless environment where system behavior is known in advance. Which is better depends on specific circumstances.

2.3.2.3   Predictability

The most important feature of real-time systems is the predictability of system behavior. System designers should know clearly all time limits in the system. It is rare in real-time systems that users can access shared data at random.

It is normal in real-time systems that many processes have already been known after the occurrence of events. Furthermore, the worst situation of these processes is also knowable. For example, if execution time of process X, process Y, and process Z is 50, 60, and 60 ms, respectively, and starting time is 5 ms, the processing time of the event can be known in advance in the system. Therefore, a real-time system must be a predictable system.

2.3.2.4   Fault Tolerance

Many real-time systems are applied to control equipment with high security. In a certain particular environment, or when there is some external interference like electromagnetic interference, communication error rate of these equipments may increase, which will affect the communication reliability of remote control systems. When key control signals are being transmitted from the proximal control room to the distal launching center, there may be serious or even devastating consequences if signal transmission is interfered by external factors. This shows that fault tolerance is very important.

2.3.3   Performance Analysis

Network and system performance is mainly reflected in the effectiveness and reliability of information transmission. Effectiveness refers to the amount of information transmission and reliability refers to the dependability of information receiving. There are some specific indexes about effectiveness and reliability.

2.3.3.1   Effectiveness (Including Transmission Rate, Bus Efficiency, and Transmission Delay)

Transmission rate is also known as transmission bandwidth, which refers to the maximum number of binary digits transmitted in unit time.

Bus efficiency η refers to the percentage of valid data transmission time t_m in the total transmission time t_0 (idle time excluded):

$$\eta = \frac{t_m}{t_0} \times 100\%$$

The ultimate service object of bus systems is valid data transmission. Bus systems have to pay extra but necessary price, including format information management information, to guarantee the reliable transmission of valid data, so bus efficiency is always less than 1.

Transmission delay τ refers to the maximum time of unit valid data transmission and it is determined by transmission rate and bus efficiency:

$$\tau = \frac{t_0}{R_b \eta}$$

In the actual transmission, message is often used as data unit, so specific standards of transmission delay are often different.

2.3.3.2   Reliability (Including System Reliability and Transmission Error Rate)

System reliability includes hardware reliability and software reliability. Hardware reliability is often measured by MTBF (mean time between failures). However, at present, software reliability cannot be measured by simple indexes. Strict quality monitoring should be carried out throughout the whole process of software programming, and long-lasting operational testing should also be conducted under various input conditions, especially under the condition of boundary input.

Bit error rate refers to the ratio of the number of erroneous code elements to the total number of code elements. It is a statistical average which reflects the impact of various interruptions and channel quality on communication reliability. Error codes do not stand for system failure, because most error codes can be detected in the process of bus error control and then be corrected or retransmitted. However, error codes indeed have negative impact on transmission efficiency. In general, error rate has relations with transmission media. Optical fibers have the lowest error rate; error rate of shielded cables is lower than that of unshielded cables.

Some other important indexes are to be introduced in brief. Transmission distance refers to the maximum physical length between two nodes on the bus. Network load refers to the maximum number of nodes mounted to a bus. Utilization rate refers to the percentage of actual amount of transmitted data in the maximum amount of transmittable data.

Performance indexes are closely related to usage occasions. For example, transmission rate and transmission distance exert mutual effects on each other, error rate is related to field interference, and transmission efficiency and delay depend on the task to a large extent.

2.4   Examples of Bus Network Construction for Testing, Launching, and Control

2.4.1   1553B for Testing and Launching Networks

At present, information flow of carrier rocket control systems is basically unidirectional; controlled parts and testing systems are in passive conditions. Subsystems work in their own way because of some traditional restrictions, which results in a huge repetitive ground system and myriad categories of equipment and interfaces with poor compatibility.

Command/response data bus architecture should be adopted due to the features of information processing in on-board control systems and relative ground testing systems. 1553B bus is the shortened name for MIL-STD-1553B, which features a dual redundant balanced line physical layer, a network interface, time division multiplexing, half-duplex command/response protocol, and up to 31 remote terminals. 1553B bus enjoys the features of high reliability, good real-time capability, and excellent fault tolerance. As a networking means for rocket control systems and ground testing systems, 1553B bus can achieve the integration of ground testing, launching, control, and comprehensive diagnosis and simplify system structures to raise system reliability and working efficiency. In addition, it can also shorten launching period, increase launching success rate, and accelerate miniaturization, modularization, generalization, intellectualization, automation, and visualization of comprehensive testing, which can improve the overall performance of systems.

Currently, all of on-board control systems and ground testing systems are device oriented, which means that devices exchange information in a point-to-point way and information output devices send information directly to information receiving devices. In bus systems, all on-board equipment and ground testing equipment should send information directly to the bus which is seen as a public channel for information exchange of the whole system. Connected with the bus through 1553B interfaces, every device mounted on the bus functions as a terminal and provides the bus with information as well as receiving information from the bus. To initialize

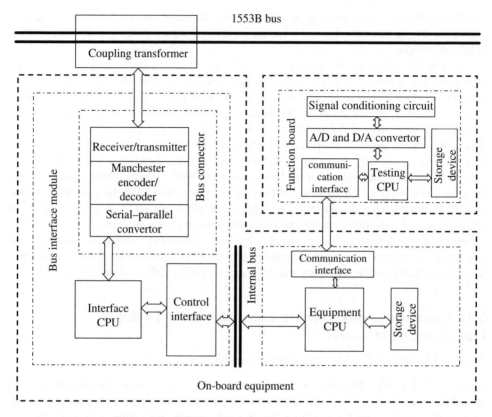

Figure 2.4 1553B bus interface and terminal structure.

the interface chip and exchange data with it, these devices should be equipped with intelligent chips, thus becoming intelligent digital devices that require Central Processing Unit (CPU), Single Chip Microcomputer (SCM), and so on.

Device intellectualization is the precondition and key technology for the design of system bus and is shown in Figure 2.4. All the following discussion is based on such precondition.

2.4.1.1 1553B for On-Board Control Systems

On-board control systems are important testing subjects of ground comprehensive testing systems. The on-board bus control system is a multilevel system that is loosely coupled in structure and tightly coupled in logic. Various sites have their own CPU and RAM and can allocate computing tasks to different sites for simultaneous computing. Each site should have a sharable local database. Such dynamic database can reduce the amount of valid data that is needed to be transmitted, shorten the computing period, and increase the density of data sampling. These are the advantages of distributed control systems.

Such flexible and reliable bus technology can not only optimize the system design and layout but also simplify on-board and ground equipment and raise the degree of automation. The whole data management system can achieve dynamic self-testing, self-diagnosing, and dynamic redundancy reconfiguration. It can use all kinds of information effectively and comprehensively, simplify testing procedures, enhance real-time fault diagnosis, and shorten the testing and launching period.

1553B bus standard is suitable for on-board control systems and relative comprehensive ground testing systems in terms of transmission mode, real-time capability, and reliability. The module of the on-board bus control system is shown in Figure 2.5.

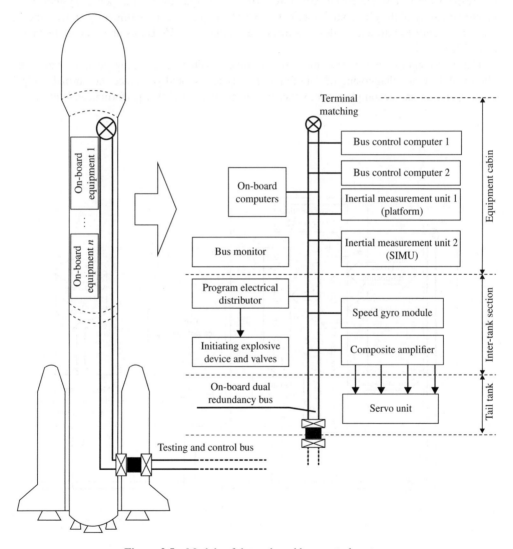

Figure 2.5   Module of the on-board bus control system.

2.4.1.2 Comprehensive Ground Bus Testing System

Shown in Figure 2.6, the comprehensive ground testing system contains the testing system, the fault diagnosis expert system, and the data management system. 1553B bus is the public data transmission channel in the system.

According to the bus standard, the ground bus control microcomputer functions as the bus controller in the whole process of testing and organizes data transmission on the bus in a specific order. Ground testing systems, expert systems, and information management computers are intelligent equipments that meet the requirement of bus interface. They have perfect functions of self-testing, data requesting, and data receiving. Through the bus, testing systems are responsible to test the rocket structure, non-electric quantity of the power system, and electric quantity of the electronic system. On-board equipment can intelligently convert these physical quantities into data codes that meet the standard of 1553B bus and then make them available for testing.

The fault diagnosis expert system is for real-time monitoring of testing data, analyzing of abnormal data, and diagnosing of data faults. The system should guarantee the normal equipment operation and accomplish comprehensive assessment of rocket performance. With mass

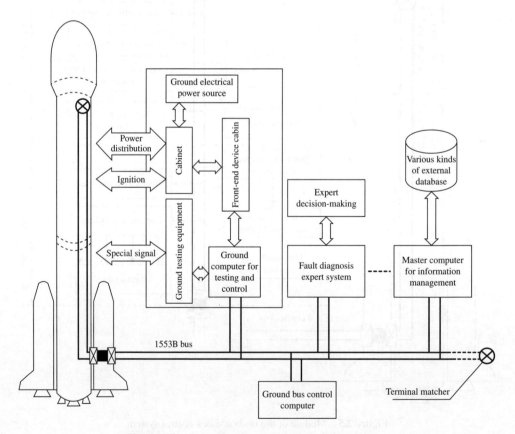

Figure 2.6 Comprehensive ground bus testing system.

storage devices, the host computer of information management is responsible for the management of empirical data, binding data, fault model data, and the knowledge base of expert systems.

2.4.1.3   Ground Testing, Launch, and Control System

In the testing, launch, and control system, front-end optical fiber data transmission devices are eliminated and replaced by the ground testing and control microcomputer which can reduce the burden of the original automatic testing system, as shown in Figure 2.7. Through communication with the bus terminal, the ground testing and control microcomputer can know testing results, so it is not necessary any more to use programmable switches to control electric relays. In the bus, the ground testing and control microcomputer can undertake digital information testing tasks which were accomplished by the VXI automatic testing system before. Moreover, with stronger self-testing capability of on-board terminal devices, ground testing will be more simplified and automatic. A majority of tests are accomplished by on-board devices automatically and original automatic testing equipments are only responsible for collection and testing of analog information.

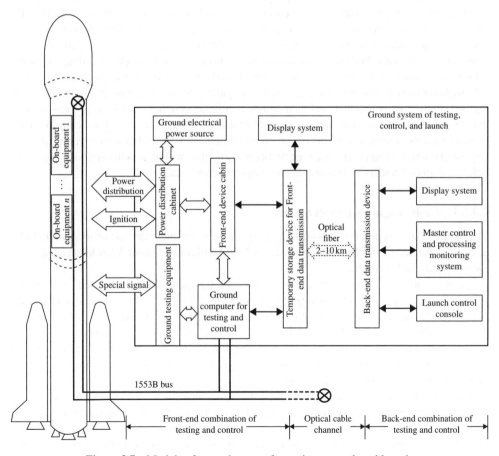

Figure 2.7   Module of ground system for testing, control, and launch.

As a terminal, the ground testing and control microcomputer is directly mounted on 1553B bus. It can not only directly receive testing information from on-board devices but also bind, check, and start on-board computers as a data transmission interface. Meanwhile, the connected automatic testing equipments are responsible for testing of on-board analog information.

For the remote testing and control combination equipment, it is optical fiber channels that link the rocket center control room and the front-end device room. The current testing and control combination technology is relatively mature, so the original testing and control combination equipment can be all retained with just testing information sources being changed. Additionally, the original way of sending signals of power distribution and ignition can also be retained.

The working process of 1553B bus is as follows.

First of all, on-board computers control terminal self testing on the bus and send the results to the ground testing and control microcomputer which will send these results to the launching and control system via optical fibers. After receiving all self-testing results and confirming them qualified, the system will start other tasks like subsystem testing and overall checking. In the process, testing results can be analyzed by programs in the center control room, or more simply be analyzed by binding programs of the ground testing and control microcomputer, and then the master control computer in the center control room will receive signals regarding whether these results are qualified or unqualified. In subsystem testing and overall checking, the on-board terminal system, controlled by on-board computers, needs to send signals according to the bus transmission protocol. During the testing, if the ground testing and control system needs testing information from a given terminal, it can make a request to on-board computers through the testing and control microcomputer, and then on-board computers will make the given terminal send data to the testing and control microcomputer. Similarly, if the ground testing and control microcomputer needs to send information to on-board computers, it should first make a transmission request to on-board computers, and then on-board computers will make a transmission command before the ground testing and control microcomputer sends data.

2.4.1.4 Fault Diagnosis Expert System

The expert system is potentially useful in the space area. An integrated expert system includes databases (the knowledge base and the question bank), inference control, knowledge acquisition module, and display interface.

A carrier rocket is a large complicated system which requires cooperation and coordination of all sections. If adding all knowledge to only one knowledge base, the knowledge base must be extremely multifarious and disorderly with low efficiency of inference and searching. Therefore, the hierarchical diagnosis model can be applied to reduce inference time because of carrier rockets' feature of structural and functional classification. Specifically, a carrier rocket is functionally classified into several subsystems, such as power distribution system, guidance system, attitude control system, and safety system. A knowledge base is then established for subsystems at the operational level. Thus, faults and errors can be located directly at the operational level and processed with the knowledge at the operational level. The knowledge in the knowledge base comes from expert experience and circuit theory analysis. In addition, combined with the comprehensive testing system, the host computer of resource management on the bus can offer large volumes of data, model parameters, and experiential knowledge for inference control and learning mechanism of expert system, thus making up for the shortfall of the knowledge base of expert system.

At present, all comprehensive testing is static testing, so promoting trend analysis of static testing data can help to get a clear picture of potential faults in advance. In the meantime, faults or errors in testing are processed and analyzed manually, so establishing expert system and eliminating faults with the help of computers can effectively shorten the testing and launching period. With historical data, expert system can comprehensively assess rocket performance and provide decision-making basis for rocket launching. The combination of 1553B bus and expert system can not only play a better role in data transmission and process but also improve greatly the real-time capability of diagnosis effect of expert system.

In the first place, data transmission of the distributed real-time control system of 1553B bus has the features of high efficiency, information sharing, and parallel computation, which can meet the requirements of the comprehensive testing system. In the second place, the distributed real-time control system's highly reliable hardware and perfect protocol processing methods can further improve the reliability of testing systems. To simplify testing procedures, bus system must have perfect software and hardware; on-board and ground bus controllers must have excellent control programs and reasonable network layout; all intelligent terminals must coordinate with each other. Carrier rocket systems and equipments are very complicated and many software and hardware problems are needed to be solved, so there is still a long way to achieve large-scale engineering application. Fortunately, 1553B bus has brought revolutionary changes to the new generation of carrier rocket and its comprehensive testing system.

2.4.2   LXI for Testing and Launching Networks

The testing, launching, and control system is required to have good performance (high accuracy, fast speed, high capacity, and multiple functions), high reliability, low cost, short development period, and excellent self-testing and self-diagnosis capabilities. Currently, the testing systems of CAMAC (Computer automated Measurement and Control) and VXI (VME Extensions for Instrumentation) are mainly adopted for on-board testing. A prolonged test proves that these two testing systems can meet the requirements of current testing projects in terms of testing accuracy and reliability. More importantly, the timely introduction of new network and testing technology is conducive to the establishment of more flexible and multi-functioned testing systems.

As to the testing base, the total perimeter of the testing site is about dozens of or even hundreds of kilometers. There are numerous testing projects, such as the testing of speed, acceleration, impact force, power, frequency, and ground telemetry testing as well as the monitoring of environmental parameters like temperature, humidity, air pressure, wind power, and wind direction. In a word, the number of monitoring points is large and they are decentralized. Under such circumstances, it is obviously impossible to adopt centralized measuring and control systems. Therefore, distributed and networked testing systems should be taken into consideration.

LXI (LAN-based Extensions for Instrumentation) bus which can reduce the size of testing modules will improve the testing function greatly. Secondly, the removal of proprietary interfaces can reduce complexity of electrical connection. Thirdly, common hardware and proprietary software in systems can raise testing flexibility and prolong the useful life of the whole testing system.

The on-board testing and launching system is very complicated, and numerous testing projects need to be accomplished. Sources and processing methods of signals under test are various, and testing data should have high accuracy. LXI bus is just designed for higher testing accuracy and efficiency on the basis of meeting the requirements of technical indexes. The LXI testing network is practically significant to promote system matching and extension capability and to improve real-time capability and efficiency of testing software.

2.4.2.1   System Testing Scheme

Take the testing of the rocket flight control system. The traditional centralized testing structure is shown as Figure 2.8.

The main feature of the structure is that the testing system centers on a master computer whose commands make testing equipments work normally. The system is highly reliable, but its devices have poor universality. Data transmission in the system is in a point-to-point way, so the amount of information is limited. Additionally, the restriction of connecting cables is not good for establishment of large-scale long-distance testing networks.

The testing system scheme is shown as Figure 2.9. In the scheme, the network-based LXI bus is utilized as the main bus with VXI, GPIB, and other buses working together. In the bus

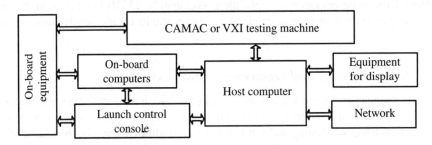

Figure 2.8   Principle scheme for structure of traditional on-board testing system.

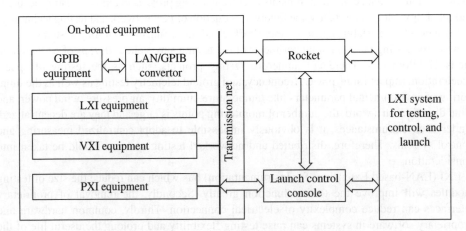

Figure 2.9   Principle scheme of LXI bus-based testing network.

network, the majority of testing tasks are completed by LXI devices. Each LXI device has its own processor, so all LXI devices can be directly connected into the network. Some certain testing tasks are completed by relatively mature VXI bus and GPIB bus which are also connected into LXI bus because of the factors of stability and cost through a LAN/GPIB converter and a controller that supports network transmission.

Compared with traditional testing bus, the LXI testing system works in a networked mode with servers and clients, which is beneficial to solving the problems that the traditional testing system module is unitary and that testing projects, testing management, and testing personnel are decentralized. A master computer works as the server, and testing projects are the testing clients. The sever can manage and coordinate testing clients in a centralized way by giving commands; testing clients will independently complete testing tasks synchronously or asynchronously after receiving commands from the server. Testing data is processed by clients and then transmitted to the server for further processing.

2.4.2.2   Strategies for Synchronous Testing

Synchronization is a basic requirement to be met during on-board testing and fault diagnosis. VXI devices can make synchronous testing available through backplane bus, but this method is infeasible for modules in different cases though feasible for modules in the same case. LXI devices contain three types of synchronous triggers: network message trigger, IEEE-1558 clock synchronization trigger, and LXI trigger bus.

Network Message Trigger
System structure of network message trigger is shown as Figure 2.10. Multiple LXI devices are linked together by a switch or a hub. Network-triggered messages are transmitted to all devices by a computer or to other devices by one certain device. Thus, the PTMP (point to multi-points) trigger can be realized. Triggered messages are transmitted in accordance with

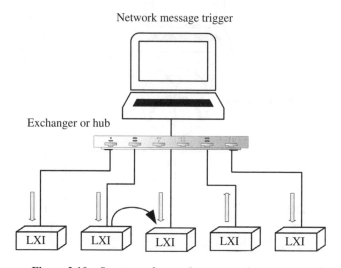

Figure 2.10   Structure of network message trigger system.

standard UDP protocol and network handshake is not necessary, so time delay of UDP protocol is lower than that of TCP/IP protocol. Such trigger mode has its own advantages: it is not necessary to use a trigger bus, it is more flexible than software trigger, and it is not limited by distance. However, the instability of UDP protocol may affect transmission accuracy. Accordingly, UDP protocol is more suitable for intermediate-and-long-distance testing systems in which data accuracy is not seen as important.

IEEE-1558 Clock Synchronization Trigger

The topological structure of IEEE-1558 clock synchronization trigger is shown in Figure 2.11. One of the LXI devices is selected as the master clock and other devices are slave clocks. The principle of synchronization is shown in Figure 2.12. The master clock sends a synchronous information packet to all slave clocks; slave clocks receive the packet and then send relative delay request information packet back to the master clock. The master clock then sends delay response information packet. With formulas, the deviation between the master clock and slave clocks can be calculated, and thus every clock can correct its own time. In this mode, triggering signals are used to tell every device when to start output signals, and each device starts at the scheduled time rather than when receiving commands from Ethernet. Therefore, network overhead and delay play no impact on event trigger. This mode is particularly suitable for long-range distributed synchronous data collection, because it is not necessary to connect cables and this mode has no distance limitation.

LXI Trigger Bus

LXI trigger bus, a M2LVDS bus, is utilized in level-1 modules. LXI modules can be configured as trigger signal sources or receivers; trigger bus interfaces can be configured as wired or logic. Each LXI module is furnished with input and output connectors which can make

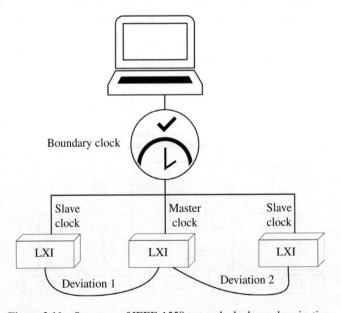

Figure 2.11 Structure of IEEE-1558 network clock synchronization.

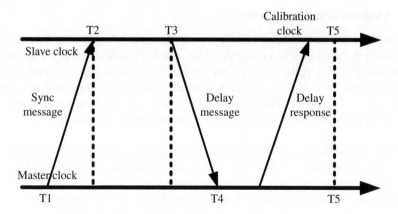

Figure 2.12 Principle scheme of 1588 clock synchronization.

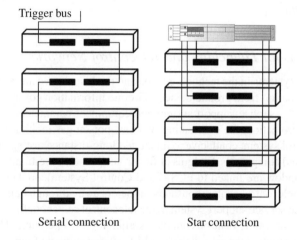

Figure 2.13 Application method of LXI trigger bus.

modules in a daisy-chain connection. LXI trigger bus is very similar to VXI and PXI backplane buses that have two connection types: serial connection and star connection, as shown in Figure 2.13. Such trigger synchronization method makes full use of advantages of VXI and PXI trigger buses. Synchronization accuracy (about 5 ns/m) is determined by the length of the trigger bus. LXI trigger bus is suitable for systems whose testing devices are close to each other.

In conclusion, synchronization accuracy of network message trigger, IEEE-1558 clock synchronization trigger and LXI trigger bus increases by their turns. Synchronization accuracy of trigger bus is about 5 ns/m; clock synchronization accuracy of 1558 network is less than 100 ns. Carrier on-board testing should have high accuracy and discarding trigger bus can reduce complexity of system connection, so it is better to use IEEE-1558 clock synchronization trigger.

2.4.2.3   Methods of Reducing Delay

LXI devices are linked to testing computers via network cables, so the data transmission distance is much longer than that of GPIB devices and VXI devices. However, testing delay is a big problem of LXI devices. Usually after a computer gives a testing command, it will cost 70 μs or even 1 ms for LXI devices to give response. The response time is determined by the speed of network handshake. The space testing system must have high real-time capability. The following methods can be used to reduce network transmission delay.

1. SCPI (Standard Commands for Programmable Instrumentation) commands directly make programmatic control of LXI devices, which can increase rate because parameters need to be converted into SCPI commands when upper drivers are used.
2. The communication mechanism of LAN sockets determines that it is better to use big data packet for each network communication and to reduce the frequency of data packet transmission. During communication with LXI devices, a series of commands can be sent to LXI devices at a time, and then one of these commands can be used to let devices start to execute the series of commands. This method can reduce delay caused by multiple times of command sending.

2.4.3   Field Bus for Launching and Telecontrol Networks

Logistical support systems in the technology area and the launch area are responsible to collect and transmit information from ground devices. Ground information covers workflow and fault handling plans of the service support system and equipment parameters, such as fuel filling, swing beam, fire protection, the collimating window, poison gas monitoring, gas supply, and UPS. The telecontrol system, conducive to integrated long-distance testing, achieves to make a telecontrol of ground equipments. Systems of fire protection, air conditioning, swing beam, and the collimating window are linked to FCS (Field Control System), so data on the bus can be transmitted to the telecontrol center thousands of meters away through optical fibers. The control computer of the service support system in the telecontrol center configures the connectivity of system signals and sends control commands after logical operation. These control commands are to be transmitted through the bus to PLC (Programmable Logic Controller) of each system which executes these commands by controlling contactors, valves, and motors.

As shown in Figure 2.14, the telecontrol system has three layers: the field layer, the control layer, and the management layer. The field layer and the control layer, whose design concept is similar to that of industrial control system, have a close control system in the control site. The management layer contains the remote testing system and the telecontrol command center that are linked to controllers like PLC through Ethernet, depending on the ground service information system and remote networks. The three layers make the control system intelligent, networked, and digitalized.

The telecontrol system in the telecontrol room can achieve full-automatic computer control. The close control system in the field equipment control room can achieve semi-automatic process operation and full-manual operation. Full-automatic computer control is more effective when everything goes well; the close control system can be put into use when there are some problems with the telecontrol system.

All condition parameters of analog quantity and switch quantity in the whole system should be connected to PLC for telecontrol and monitoring. All controlled subjects must be able to be

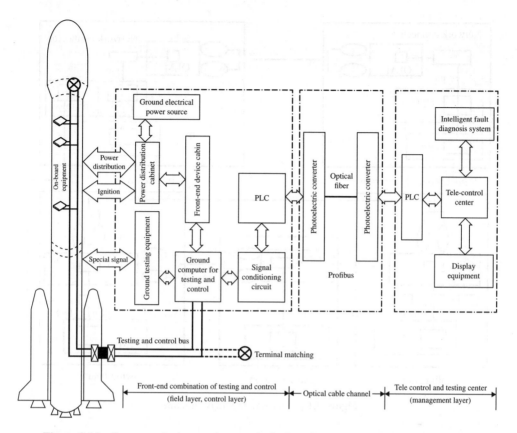

Figure 2.14 Structure of telecontrol system in the launch area with telediagnosis function.

controlled by both telecontrol IPC (Industrial Personal Computer) and close control computers; all state signals must be able to be displayed on the screens of both telecontrol IPC and close control computers to make the telecontrol system more reliable and flexible.

It can be seen that the distributed telecontrol system in the launch site has made use of various types of technologies, such as computer control technology, industrial network technology, PLC control technology, and field bus. Such technological combination can not only protect equipment and technology investment and reduce the cost of equipment modification, but also well integrate the field control network with the Ethernet-based network.

The design, establishment, and modification of the control system are conducted at the management layer above the control layer. Control technologies below the control layer are mainly mature industrial control technologies.

2.4.3.1 PROFIBUS-DP Field Bus Network

PLC of the remote station and PLC of the short-range master station are linked by optical fibers as a field bus control network; three I/O stations are linked with DP by OLM (Optical Link Module) as a subsystem control network. The PLC master station has industrial Ethernet interfaces which can offer timely accurate condition parameters to the ground service server and

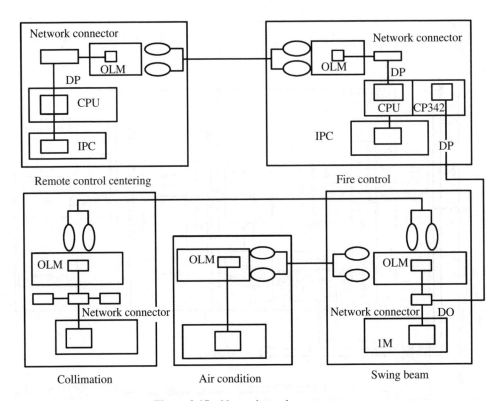

Figure 2.15   Network topology structure.

make field debugging easier as reserved interfaces that support long-distance technologies. Network topological structure is shown in Figure 2.15. PLC is used in the control system in which there is a PROFIBUS-DP network. Optical cables are utilized in bus control to prevent impulse interference. The upper computer, linked with CPU through MPI interface, finishes system configuration and program installation firstly; following program installation can be accomplished through MPI or industrial Ethernet. Fire protection equipment, swing beam equipment, collimating window equipment, and air conditioning equipment, as slave stations, are connected onto DP bus of front-end PLC. Three DP networks connect bus equipment into the industrial control network through DP/DP connector, which makes "centralized management" and "decentralized control" come true. Telecontrol station can be located in the telecontrol center of the technology area. Internal PLC and the PLC in the control station of the launch are can make telecommunications and send control signals to each other. Host computer monitoring enables the commander of the ground service support system and telecontrol operators to check feedback signals and parameters on the display screen.

2.4.3.2 Application Analysis and Prospect

The PLC network formed by the PROFIBUS field bus can raise control capability and reliability of the control system and reduce operation cost. It has the following advantages.

Easy for Modification and Maintenance
In the configured automatic system, sensors and actuators are linked by cables that are plugged directly into the I/O board of the central PLC. Usually, numerous cables are needed, especially for some work sites far from the control center. It is advantageous to place the I/O board close to field sensors and actuators and link the board with sensors and actuators through DP.

Information Integration of the Control System
Configuration is convenient; programming of control programs is simple. Field bus systems based on distributed configuration are not quite different from systems based on traditional central configuration, but the design of control software and monitoring system configuration in field bus systems is featured by simplicity and rapidity. Input variables from sensors linked with distributed I/O and output variables from actuators linked with distributed I/O are considered as I/O of the central PLC to be processed, so communication programs are not needed any more. The design of human–computer interface can just simply define and process field data and former control programs can be transplanted simply.

Being more and more intelligent, digitalized, information-based, networked, miniaturized, and decentralized, field bus control systems featured by bidirectional serial digital multiple-node communication will definitely replace traditional control systems such as direct digital control systems based on central controllers or DCS (Distributed Control System) characterized by hierarchical controllers.

Applying filed bus technology as well as equipment's computing and information processing capabilities into the ground equipment self-control system can improve the advantages of integration and comprehensiveness. Field bus digital measuring instruments will replace analogue measuring instruments gradually. FCS will modify the traditional DCS structure step by step and then replace it completely. In the transitional period, integration of FCS and DCS, the way which must be taken, is consistent with the actual situation of ground equipment establishment and is beneficial to application and development of FCS. However, current ground equipment self-control system is focusing on DCS, FCS just being a constituent part. Accordingly, the performance of the whole system, like reliability or openness, can be improved further.

It is necessary to further study the application of FCS products, software development, and integration of FCS and DCS for ground equipment establishment of field bus. That FCS is applied more effectively in the future will definitely enhance information construction of ground equipment in the launch site.

2.5   Transmission Networks of Command Systems for Spacecraft Testing, Launching, and Control

2.5.1   Overall Architecture of Transmission Networks

Transmission network of the control and command system for space launch is a hierarchical and modularized architecture with three layers: the core layer, the convergence layer, and the access layer. The core layer provides a backbone transmission and switch network which has strong switching capacity and various types of network interfaces and protocols. Such network is highly reliable, stable, and extensible. In the middle is the convergence layer linking the

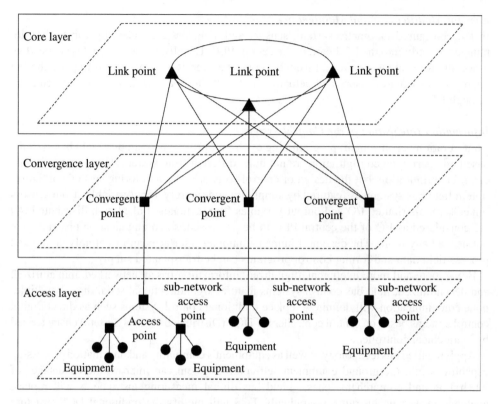

Figure 2.16   Overall architecture of transmission network.

upper core layer and the lower access layer. Having convergent points for multiple switches of the access layer, the convergence layer can process all communication information from the access layer and transmit them to the uplinks of the core layer. The access layer has terminal user interfaces that enable terminal users to be linked to the transmission network, and the access layer can link all types of users to the convergence layer. The overall architecture of transmission networks is shown as Figure 2.16.

Network configuration of each layer depends on functional partitioning and task undertaking. The network architecture is shown in Figure 2.17.

2.5.1.1   The Core Layer

As a backbone network, the core layer is the center of the internal network of the launch center. Linking networks of the launch center and nodes of the convergence layer, the core layer is designed to guarantee unimpeded and high-speed data exchange and fast route convergence. Usually functioning in important places such as the command center, the testing and launch center, or the communication center, the core layer is required to have strong switching capacity, various types of network interfaces, and protocols as well as high reliability, stability, and extensibility. The core layer equipments of the transmission network of the control and

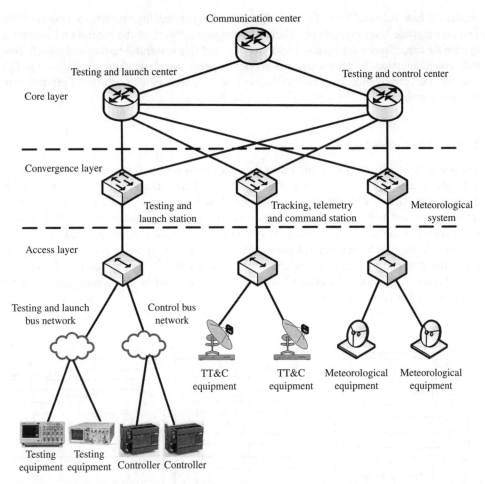

Figure 2.17 Transmission network architecture of testing, control, and command system for space launch.

command system for space launch compose of three network nodes—the command center, the testing and launch center, and the communication center. Each node is equipped with highly reliable and high-performance 10G Ethernet switching equipments which link the three nodes as a loop circuit.

2.5.1.2 The Convergence Layer

Linking the core layer upward and the access layer downward, the convergence layer is a regional center for the Intranet of the launch center, mainly responsible for information convergence and exchange in a given region. Nodes of the convergence layer are often deployed in units with large amount of data. As an implementation layer of QoS strategies such as bandwidth limitation, flow distribution, and access control, the convergence layer is required to have balanced loads, fast convergence, and high reliability, stability, and extensibility. The convergence layer equipments of the transmission network of the control and command system for space launch are used in important places of five systems—testing and launch, control, communication, meteorology, and service support. As a switching center for the five systems, the convergence layer adopts double-route connection with additional protection of the egress router.

2.5.1.3   The Access Layer

The access layer is edge device and offers network access for the five systems. Used in many tasks, the access layer has functions of user access authentication, layer 2 switching, VLAN division, and data distribution. The access layer should have easy access, low environmental requirements, and high performance at a low cost. The access layer equipments of the transmission network of the control and command system for space launch are used in tasks of the five systems—testing and launch, control, communication, meteorology, and service support. It guarantees the front-end equipment access and provides users with 100 M or gigabit broadband.

As Figure 2.18 shows, the network architecture of the control and command system for space launch has come into being.

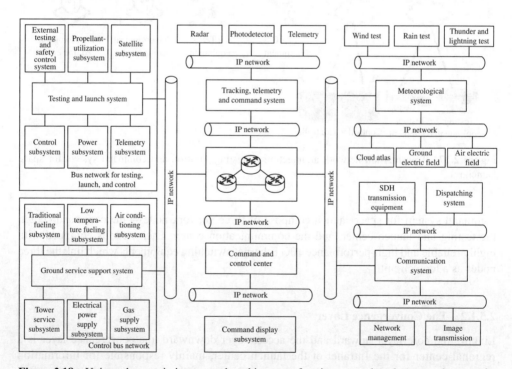

Figure 2.18   Universal transmission network architecture of testing, control, and command system for space launch.

According to Figures 2.17 and 2.18, the transmission network of the control and command system for space launch has the following features.

1. It is a triangle looped network which is more suitable for alternative routing and can realize fast routing convergence.
2. Three layer cascade switching structure from the core layer to the access layer can reduce transmission time to a large extent.
3. Simple and clear, the network architecture is beneficial to fault isolation and management. Each layer has its own functions and is easy to maintain.
4. TCP/IP protocol applied in the network is conducive to interconnection in the entire network.

2.5.2   Reliability of Transmission Networks

Data transmission of the control and command system for space launch has exceedingly high requirements for network reliability. The data transmission network in tasks must be a hundred percent reliable, and data transmission faults should by no means occur due to network anomaly. Transmission network reliability of the control and command system for space launch involves hardware reliability, link reliability, and networking reliability.

2.5.2.1   Hardware Reliability

Switching equipments of the core layer and the convergence layer should be exceedingly reliable, so it is better to use high-end switching equipments with strong switching capability and high reliability. These equipments are expected to support dual master control, power redundancy, cooling fan redundancy, distributed transmission, and hot-plug operation. If any unit breaks down, the system should switch over automatically to guarantee high reliability. Hardware reliability of a single unit should reach 99.999%. Moreover, the probability of operational errors can be reduced through reducing the complexity of core equipments.

The hot standby of switching equipments can also raise the system reliability, because the fault system can be recovered timely with the security of the hot standby. Switch between master and slave boards is manipulated by applications of main control boards. There are two main control boards in the system: one is the master board which controls the whole system and the other is the slave board which serves as a backup. The main control board, through the hot switching controller, functions to detect hardware faults or plugging and unplugging operations that may bring negative results to the system. The two main control boards can exchange heartbeat data with each other through Ethernet interfaces. When the salve board is detecting faults of the master board, switch between the two boards should start and the system bus should be connected and controlled. Then the former master board will work as a slave board after it returns to normal. Figure 2.19 shows the theory.

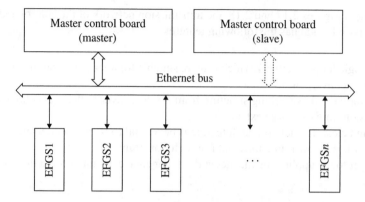

Figure 2.19   Principle of master control backup.

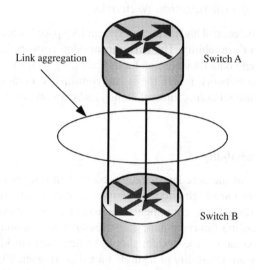

Figure 2.20   Example diagram of Eethernet port trunking.

2.5.2.2   Link Reliability

Port trunking refers to a port set which can increase throughput beyond what a single connection could sustain and provide redundancy in case one of the links fails (shown as Figure 2.20). Port trunking has three types—manual aggregation, dynamic LACP (Link Aggregation Control Protocol), and static LACP. All ports should be consistent with each other in one port set, which means that if one port is Trunk, other ports should be Trunk as well; if the port's link type is Access, the other ports should have the same link type. Basic configuration of ports includes STP, QoS, and VLAN, etc.

LACP based on IEEE802.3ad is a protocol for dynamic link aggregation and disaggregation. LACP works by sending LACPDU (Link Aggregation Control Protocol Data Unit) down all links that have the protocol enabled. After a certain port has LACP enabled, the port will send LACPDU to inform the other end of its system priority, MAC address, port priority, port number, and operation key. After receiving these information, the receiving end will compare these information with that from other ports to select aggregatable ports. Thus, the two ends can reach an agreement on joining or quitting some dynamic port set.

2.5.2.3 Networking Reliability

When Ethernet technologies are applied in the network, the cascade connection of switches linking the core layer and the access layer should be no more than four levels, because this can reduce the convergence time. When it comes to the establishment of STP ring protection, triangle rather than quadrangle or polygonal looped network is more appropriate and reliable. In addition, three levels of switches linking the core layer and the access layer can make more reliable the transmission network of the control and command system for space launch.

2.5.3   QoS Solutions for Transmission Networks of Control and Command

IP-based networks are shared networks. Their "best-effort" transmission mode is in no position to secure end-to-end quality of service. Network index and parameters will change dynamically with the change of transmitted data, which makes IP QoS become a design difficulty. At present, there are multiple IP QoS technologies and models, but not a single one of these can meet all demands and requirements. Therefore, a combination of different technologies is needed and matching QoS strategies should be applied in different aspects of networks. Only in this way can network resources be used effectively and quality of service be guaranteed.

Data transmission of the control and command system for space launch must be characterized by high QoS. Based on the feature that network traffic flow and its directions are predictable, network monitoring and management should be greatly enhanced and QoS problems should be effectively solved with the application of comprehensive technologies. Specific solutions are as follows.

1. Ten-gigabit Ethernet technology should be used to establish the transmission network of the control and command system for space launch. The broadband with a wide and light load must be ensured. QoS requirements, such as time delay and packet loss rate, must be met to improve quality of service.
2. MPLS (Multiprotocol Label Switching) should be enabled among equipments of the core layer in the transmission network of the control and command system for space launch. A MPLS domain then comes into being. In the whole MPLS domain, labels bound to data packets determine the forwarding priority of data packets. For data packet transmission, packets in the queue with high priority are always transmitted first, and packets in the queue with low priority are transmitted in the second place. Data packets with high priority will never be lost even if the network is congested.
3. In the transmission network of the control and command system for space launch, equipments of the convergence layer should be classified according to their priority level. Many QoS strategies should be implemented, such as broadband limitation, access control, traffic shaping, traffic supervision, and queue scheduling. Packets with high priority must always be processed first to meet the QoS requirement.

4. VLAN (Virtual Local Area Network) should be applied to divide the control and command network for space launch. Network broadcast must be restricted to each logical subnet, or network broadcast storms cannot be avoided.

5. Audio, video, and large file transmission must be restricted in the control and command network for space launch. Task networks are only allowed to transmit task data. Task images and voice can be transmitted by other communication modes.

6. In task programming, the multicast mode which can reduce data flow will gradually replace UDP broadcast.

7. It is a must to limit strictly the number of users and broadband of user that have access to the control and command network for space launch. Unnecessary users are not allowed to have access to the network, which is beneficial to data flow decrease.

2.5.4 Transmission Network Security

Involving many aspects, network security is a comprehensive, complicated, and developing issue. There exists no such solution that can get the issue addressed once and forever. Enough attention should be paid to the issue, and management skills and scientific technologies should be improved constantly. Internal LAN with optical fiber connection is adopted in the whole launch area except the distant measuring station that uses WAN. Information is only exchanged within the launch center. If information cannot be transmitted outward, network environment will be more secure and reliable. As shown in Figure 2.21, security of the private IP network for control and command tasks in space launch is designed as a hierarchical structure, including physical security, link security, network security, host computer security, application security, and data security. Figure 2.22 shows the security protection architecture for the transmission network of the control and command system for space launch.

2.5.4.1 Physical Security

Electromagnetic shielding and access control systems are used in the central computer room of the transmission network to guarantee physical security of all network equipments, because they can shield electromagnetic radiation and prevent leakage of important

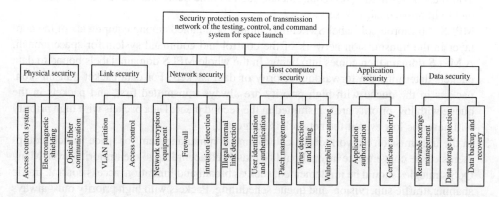

Figure 2.21 Security protection of transmission network of the testing, control, and command system for space launch.

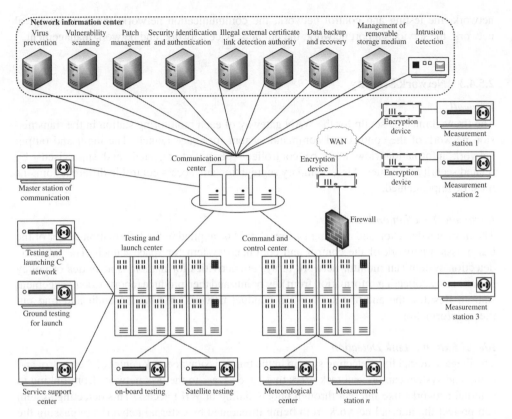

Figure 2.22   Security protection system structure of transmission network of the testing, control, and command system for space launch.

information. In the task network, optical fibers are used to link the core layer, the convergence layer, and the access layer, because optical fiber communication has less electromagnetic radiation which can prevent information leakage during transmission.

2.5.4.2   Link Security

Hubs are forbidden in the transmission network of the control and command system for space launch, because their mode of shared data transmission may result in undesired data interception.

VLAN technology has been used to divide the transmission network of the control and command system in an effective way. Such technology can separate different users and applications into corresponding virtual subnets so that VLAN data is only transmitted within its own virtual subnet, which will definitely promote data security. Additionally, layer-2 access control lists can realize flexible controlled accessibility among all virtual subnets. According to access requirements of the control and command system for space launch, the whole network should consist of several virtual networks including the carrier on-board testing network, the satellite testing network, the testing and launching C³ network, the testing and launching ground network, the testing and control network, the communication network, the meteorological network, the service support network, the public service network, and network management.

2.5.4.3 Network Security

Firewall

The firewall must be set up for the WAN gateway of each monitoring station in the transmission network of the control and command system for space launch. The input and output network information flow must be controlled—permitted, rejected, and supervised—by related security strategies. Also, boundary segregation and access control in different domains must be implemented.

Intrusion Detection and Audit

The intrusion detection and auditing system should be applied in the information center of the transmission network of the control and command system for space launch. The intrusion detection system can monitor network or system activities for malicious activities or policy violations by collecting information from key points and then produce reports to a management station. Besides, the auditing system can conduct network information audit and find out illegal information.

Illegal External Link Detection

The illegal external link detecting system in the transmission network center of the control and command system can prevent task users that have access to the task network from linking to external networks like Internet through phone dialing or WIFI access. This detecting system can protect the internal network from being threatened by external networks bypassing the firewall.

Network Encryption Equipment

Network encryption equipments are supposed to be set up for WAN connection between the command center and the monitoring station, which can ensure security and confidentiality of task information transmission.

2.5.4.4 Host Computer Security

Antivirus System

The comprehensive antivirus system in the control and command network for space launch can prevent network virus from making damages to network and user terminals. Antivirus software should be installed in all user terminals, and the antivirus server is responsible for automatic upgrade of virus database. In general, the system works as a strong protector against virus for the whole network.

Vulnerability Scanning

The vulnerability scanning system in the transmission network center of the control and command system, consisting of the vulnerability scanning server and the console, can conduct vulnerability scanning for user terminals and servers in the network on a regular basis.

Detecting and addressing security vulnerabilities timely, the system can improve network security in a dramatic way.

Patch Management

The patch management system in the transmission network center of the control and command system consists of the security analyzer installed on the client side and the automatic patch upgrading system installed on the server side. The client side is supposed to be compared to the database of the server center in a regular basis to find out necessary system patches and then download and install them automatically in case of security loopholes.

User Identification and Authentication

Due to the rigorous user identification and authentication mechanism in the transmission network of the control and command system, only authorized users can have access to the network. In the identification and authentication mechanism, network security strategies are enforced on the user side, network using behavior of terminal users is strictly controlled, and active defense ability of terminal users is enhanced to guarantee network security. The user identification and authentication mechanism consists of four parts—the security strategy server, the security client (authentication software), security interaction equipments (network switches), and the third-party server (the patch upgrading server or the antivirus server). Specific identification and authentication process is as follows.

1. After the terminal user gets access to the network, the security client will upload the user's information (including username, IP address, MAC address, VLAN number, and switch port) to the security strategy server that conducts user identification and authentication. Illegal users will be deprived of access to the network.
2. Legal users are required to get further security identification and authentication. The security strategy server will verify the client-side system patch version, the antivirus software version, and whether or not the user is infected with virus. Unqualified users will be isolated into a special network isolation region by the security interaction equipment.
3. Users in the isolation region, according to given security strategies, can get system patches installed and virus database upgraded through the third-party server.
4. Security qualified users will implement security strategies (the ACL access strategy, the QoS strategy, enable/disable proxy service) from the security strategy server. Security interaction equipments will provide network service based on user authorization, so users' network behavior can be restricted under the user authorization.

2.5.4.5 Application Security

Certificate Authority

A certificate authority center needs to be set up for the transmission network of the control and command system for space launch. The certificate authority center can provide task users with digital certificate, electronic signature, electronic seal, and other applications, in order to guarantee privacy, completeness, and authenticity of classified information during transmission and to protect information from being stolen, tampered, and forged.

Application Authorization
All kinds of application systems in tasks have the function of hierarchical authorization, which means the application system will grant permissions according to the user importance. Users are only permitted to read authorized information and unauthorized users have no access to the system.

2.5.4.6   Data Security

Data Backup and Recovery
The online data backup and recovery system for the transmission network of the control and command system for space launch adopts redundant array of inexpensive disk (RAID) and RAID-5 storage mode that can raise reliability of data storage. Once data loss or data corruption occurs, the online data backup and recovery system will restore task data in time.

Storage Protection
Important classified servers and disk arrays in the transmission network of the control and command system must be encrypted. Combined with certificate authority, disk encryption, file, and database is an effective way to protect storage of classified data.

Management of Removable Storage Medium
The management system of removable storage medium in the transmission network of the control and command system has the functions of user registration, virus defense, log audit, etc. In such system, the removable storage medium can be managed and controlled strictly. Only after registration can the removable storage medium in task networks of the launch center be used; without registration, the removable storage medium cannot be identified on task computers.

2.5.5   Network Administration Systems

Network administration systems are extremely important for the network establishment. With network administration systems, administrators can know the network's overall performance, connected conditions, user access behavior, flow distribution, application distribution, and fault warning in time. Network administration systems are responsible for remote equipment management, user authentication, bandwidth limitation, access control, network performance testing, and packet protocol analysis. Being qualified assistants of network administrators, network administration systems can guarantee long-term network stability and improve the ability of emergency response under unusual circumstances. The establishment of network administration systems in the transmission network of the control and command system for space launch should rest on equipment management, fault management, flow management, and performance management.

As the core and foundation of network establishment, network administration systems have the functions of equipment auto-discovery, topology management, equipment management, warning management, performance management, and resource management. They can display network topology flexibly, discover hidden dangers timely, change network configuration efficiently, and analyze network conditions dynamically. In network administration systems, management information is usually transmitted under SNMP (Simple Network Management Protocol).

SNMP, a communication protocol between the network administration system and network equipments, is a widely accepted industry standard and has been applied at a large scale. As an Internet-standard management protocol, SNMP functions to guarantee that management information should be transmitted between two specific points, so that it will be easier for network administrators to search and modify information at certain points and to complete fault diagnosis, capacity planning, and reporting.

SNMP consists of three parts—SMI (Structure of Management Information), MIB (Management Information Base), and SNMP itself. SMI operates in SNMP to define sets of related managed objects (host computers, routers, network bridges, and other network devices) in a MIB and to form a database by organizing these managed objects according to a certain structure. MIB is a database used for managing the entities in a communication network. SNMP itself exposes management data in the form of variables on the managed systems, which describe the system configuration. These variables can then be queried or sometimes set by managing applications.

Increasingly wide application and expanded scale enable networks to provide more services. Mastering network services and net traffic features to optimize bandwidth allocation is one of the biggest challenges for network administrators. Network traffic monitoring systems, based on flow information statistics and analysis, can discover network bottleneck and abnormal phenomenon timely by collecting and analyzing information of net traffic and resource usage. Network traffic monitoring systems can also withstand virus attacks and offer enough scientific foundation, based on traffic data of different services and applications, for fault diagnosis, network optimization, and network expansion.

NTE (Net Traffic Exporter) is responsible to analyze network traffic, collect traffic statistical information, and then transmit the statistical information to NTC (Net Traffic Collector). NTC is responsible to analyze the messages from NTE and then integrate the data into a database for the analysis of NTP (Net Traffic Processor). NTP is responsible for data extraction and statistical analysis, providing data basis for all services.

Network performance will change with many factors, such as network combination and the increase of new technologies, new equipments, and new users. It is necessary to get a clear picture of network performance to make the network stable and to make timely adjustments. The network pressure measurement system is to give different measurement instructions to different private node computers in the network through the console. These private node computers will then generate simulated traffic and analyze performance indexes of the traffic like response time, throughput rate, and effective bandwidth which show whether or not the network is in good condition. The network pressure measurement system can be used to conduct pressure measurement of simulated traffic at regular intervals. Such regular human-conducted traffic measurement can help people to know whether the network system is in good operation, to get accurate indexes and parameters that influence network QoS, to analyze hidden faults, and to avoid and reduce system failures.

3

Intelligent Analysis and Processing for Testing Data

The testing, launching, and control system for space launch and its related networks have made integration, sharing, and interaction of testing data possible, and offered support to intelligent analysis and processing for testing data of different components, subsystems, and the overall systems. Processing, comparing, and analyzing testing data are key tasks that can ensure the success of space launch and the important foundation for system performance evaluation, fault analysis, and diagnosis.

The large scale and complexity of present carrier rockets have led to increasing testing points, testing data, and data types, which may influence system security, reliability, and robustness. Traditional methods and strategies can no longer meet the demands of the present and future testing development. Artificial intelligence theories and methods, however, can well solve many problems, such as modeling analysis, nonlinear numerical calculation, and information expression in complex system. Because of these advantages, intelligent analysis and processing has become the research focus in the area of testing data analysis and processing.

In this chapter, the author will elaborate intelligent analysis and processing methods for testing data on the basis of wavelet analysis, cluster analysis, rough set analysis, and kernel function, focusing on the features of time varying, multiscale, nonlinearity, and dynamism that testing data have. Also, the author will discuss antinoise measures, singular point detection, consistency analysis, integrity analysis, and association analysis for testing data in collection and transmission.

3.1 Contents in Space Launch Testing

3.1.1 Assignments in Space Launch Testing

Testing assignments are as follows.

1. *To assess product performance comprehensively* Whether product function and performance index meet the requirement of the design and whether the manufacture

---

Intelligent Testing, Control and Decision-Making for Space Launch, First Edition. Yi Chai and Shangfu Li.
© 2015 National Defense Industry Press. Published 2015 by John Wiley & Sons Singapore Pte Ltd.

and processing meet the requirement of technology must be verified, or products should not be used for space launch.

2. *To locate and clear product failure timely* After fault location and mechanism analysis (failure property, cause, and location), certain measures should be taken to clear product failure. The design and manufacture technology needs to be promoted according to results of failure analysis.

3. *To formulate testing projects and design testing status and methods* Testing status should, to the largest extent, simulate the process of rocket flight, with rigorous methods, genuine states, and complete projects.

4. *To forecast accidents and work out an emergency response plan* Product testing has strict time requirement. The time delay caused by failure clearing in a system will impact the whole system in a negative way. Therefore, it is a must to forecast accidental failure, enhance advance coordination, and work out a response plan for potential accidents or problems before testing, which can be avoided and the failure that has happened can be coped with correctly and timely.

3.1.2   Contents

3.1.2.1   System Testing

According to various functions of testing subjects, system testing includes control system testing, power system testing, telemetry system testing, off-ballistic trajectory measurement and security system testing, and satellite testing. The feature of system testing is to use distinct testing methods and formulate testing projects for different system functions and performance requirements.

Control System Testing
This is a complicated and precise system, which consists of guiding system, attitude control system, power supply system, and program instruction system, controlling the process of rocket flight automatically. Control system testing, an important part of product testing, is conducted from one single machine to a system, which means it is conducted in a hierarchical way from unit testing at the stand-alone level to comprehensive testing at the system level.

Power System Testing
Power system includes the engine, the propellant storage tank, delivery system, pipelines, valves, and electric detonators, etc. Consisting of the gas circuit, the liquid circuit, and the electric circuit, power system is a complicated system that powers carrier rockets. Propelling system testing is also an important part of product testing. Major testing projects include gas tightness examination, testing, and installation of initiating explosive devices, pressurization of the storage tank, and so on.

Telemetry System Testing
Telemetry system is for the measurement of operating state parameters and environmental parameters. The system can help to assess product performance, analyze faults and errors, and improve design and manufacture technology. As an imperative part of product testing,

telemetry system testing includes power supply examination, sensor examination, convertor and encoder examination, and AC–DC converter examination.

Off-Ballistic Trajectory Measurement and Safety System Testing

Off-ballistic trajectory measurement system and the security system are used to measure off-ballistic trajectory parameters and local image information of carrier rockets and are helpful in assessing product performance and analyzing faults and errors. The security system is applied to receive orders from the ground and explode faulted carrier rockets to ensure the security of launch sites and aviation zones.

Satellite Testing

Satellite testing includes single machine test, subsystem testing, and comprehensive testing before and after the final assembly, which includes power supply testing, telemetry and tracking testing, attitude and orbit control testing, communication transponder testing, and payload testing. Under the condition of unified power supply and distribution, holistic testing is to examine the electrical performance and function of satellites in the round, including consistency, coordination, and electromagnetic compatibility testing of electrical interfaces that link different subsystems. In fact, satellites, manned spacecrafts, space stations, and carrier rockets are all independent and large-scale systems which consist of various systems. The reason why they are sometimes called subsystems is that their system scale is relatively small when compared with that of the overall space system.

3.1.2.2 Hierarchy Testing

Testing of spacecrafts and their carriers hierarchically contains unit testing, subsystem testing, match testing, and sum-check testing.

Unit Testing

Unit testing equipments are used for independent testing of stand-alone devices in all systems. Unit testing is usually conducted on ground when the testing subjects are in the detached state which refers to the detachment of appearance, function, performance parameters (static parameters and dynamic parameters), and error coefficient. The focus of unit testing lies in the function and performance parameters of the complete unit. Unit testing equipments should show high precision. So should devices that decide guidance precision, such as accelerometers, gyroscopes, and computing devices. More importantly, the testing needs to be conducted in a good condition. As for the testing of attitude control devices, the focus lies in static parameters, dynamic parameters, and input–output polarity.

Subsystem Testing

Subsystem testing refers to the testing and examination of subsystems' functions, performance parameters, and coordination. The focus of subsystem testing lies in each subsystem's most representative parameters, such as static parameters and dynamic parameters of stand-alone devices and subsystems. Subsystem testing involves a large number of testing projects, and a huge amount of data needs to be collected and coped with.

Match Testing

Match testing should be carried out among all systems after subsystem testing. The main task of match testing is to examine the coordination among all systems and the working condition of interfaces linking different systems. Additionally, system match testing is conducted to verify whether the comprehensive system design is reasonable and whether the electrical interfaces between the telemetry system and other systems are matched.

Sum-check testing

Sum-check testing is a massive testing when all systems are linked as a whole, which is focused to test the coordination, major functions, and performance parameters of the whole space system. Because of different testing focuses and ground set states, sum-check testing is classified into flight and launch examination that includes emergency shutdown examination. Usually, on-rocket states are basically the same during sum-check testing.

3.1.3   Requirements in Space Launch Testing

Testing requirements are as follows.

3.1.3.1   Requirements for Testing Equipments

Product testing cannot be conducted without testing equipments whose quality and precision exert an important impact on the security and reliability of spacecraft launch testing. Required to be reliable, secure, accurate, and automatic, testing equipments should have the function of self-testing and self-diagnosis.

3.1.3.2   Requirements for Testing Sites

Testing sites should have enough space for testing of each system, adequate power and gas supply, excellent grounding system, and reliable communication and command system. Moreover, testing sites should have temperature and humidity environments that are suitable for some microcomputers and precise instruments. In a testing site, there should be a shielded room for electromagnetic compatibility experiments. Accurate geodetic coordinates, azimuth reference, and vibration isolation foundation are also needed if there are some special laboratories like the inertial device laboratory in testing sites.

3.2   Data Preprocessing

Due to the influence of testing equipment states, transmission channels, testing environments, and electromagnetic interference, testing data of the carrier rocket in the central computer are usually overlaid with a lot of nonstationary Gaussian noises and white noises. The joint-test data in many tasks indicate the traditional noise-processing methods may lead to problems that at some time periods and under certain conditions, some processed data curves are glitch and jump points that actually do not exist may appear. More than that, though, data may be distorted and some valuable signals will be overwhelmed by noises when signal-to-noise ratio

(SNR) is too low. As a result, data analysis will be inaccurate or some important data features can even hardly be obtained.

Take the inertial navigation system of carrier rockets. In practical testing and experiments, multifaceted interferences make testing results not completely consistent with the truth. For example, random noises of laser gyroscopes are made up of white noises and fractal noises. Being a nonstationary random process, the fractal noise cannot be removed easily through traditional methods. Therefore, software compensation is more essential for the improvement of practical precision. With increasingly higher requirements of data processing in testing tasks, it will definitely be a megatrend to apply Wavelet Transform into rocket data processing, which can make up for the deficiency of traditional methods.

3.2.1 Noise Features in Wavelet Transform

Traditional Fourier transform cannot make a local analysis of signals in the time domain and cannot detect and explore signal mutation. If Fourier transform is applied, noises will be regarded as useful signals and then be processed together with signals in the frequency domain. When local features of data signals are needed, real data features as well as noises have to be amplified simultaneously; similarly, local features of data signals will also be weakened if noises have to be reduced. While unlike Fourier transform, wavelet transform has the feature of time-frequency localization, so it is possible to adjust the time-domain window and the frequency-domain window as required. Accordingly wavelet transform has become an important method for data noise reduction.

Let $x(t)$ be the noise. After wavelet decomposition, the low frequency information will influence the bottommost layer and the low frequency layer in following decomposition; the high frequency information will only influence details of the first layer in following decomposition. If $x(t)$ is white Gaussian noise, with the increase of layers of wavelet decomposition, the amplitude of the high frequency coefficient will be attenuated rapidly. Apparently, the variance of wavelet coefficients has the same change trend.

Let $C_{j,k}$ be the wavelet coefficient after noise decomposition. k is a time subscript and j is a scale subscript.

Following are some of the characteristics observed when the discrete-time signal $x(t)$ represents a certain kind of noise.

1. When $x(t)$ represents zero-mean, stationary, and colored Gaussian noises, the wavelet coefficients after the decomposition of $x(t)$ also form a Gaussian sequence. As for every scale j in the wavelet decomposition, the related wavelet coefficients form a stationary and colored sequence too.
2. When $x(t)$ represents Gaussian noises, the wavelet coefficients after decomposition should obey a Gaussian distribution.
3. When $x(t)$ represents zero-mean and stationary white noises, the wavelet coefficients after decomposition are independent of each other.
4. When $x(t)$ represents noises of known correlation functions, the coefficient sequence after wavelet decomposition can be calculated according to the related correlation function.
5. When $x(t)$ represents noises of a known correlation function spectrum, the spectrum of the wavelet coefficient $C_{j,k}$ and the cross spectrum of the scale j and j' can be calculated according to correlation function spectrum.

6. When $x(t)$ represents zero-mean and fixed ARMA (Auto-Regressive and Moving Average) models, the related wavelet coefficient $C_{j,k}$ of each scale j is also a zero-mean and arranged ARMA model. The characteristic of the wavelet coefficient only relies on the decomposition scale j.

3.2.2 Wavelet Denoising Algorithms Based on Threshold selection

There are two procedures for wavelet denoising based on threshold selection.

1. An appropriate wavelet basis and decomposition layer must be decided before the wavelet decomposition. Then threshold processing of wavelet coefficients should be performed at different decomposition scales. The threshold processing of wavelet coefficients at coarser scales may eliminate important features in signals; the threshold processing of wavelet coefficients at fine scales may result in insufficient denoising. Therefore, the selection of decomposition layer is as significant as that of wavelet basis. In general, decomposition layer is $n = (\log_2 m) - 5$ and m refers to signal length.
2. A threshold value should be selected to make threshold quantization processing for wavelet coefficients at different decomposition scales.

There are two methods of threshold processing: hard threshold method and soft threshold method.

3.2.3 Selection and Quantization of Threshold

3.2.3.1 Methods of Threshold Denoising

There are two methods of wavelet threshold denoising, namely hard-threshold denoising and soft-threshold denoising. The hard-threshold denoising is disadvantageous in filtering the selected threshold during denoising, which is less precise, especially when instance variables are present in signals. While for soft-threshold denoising, the risk of threshold selection is lowered because reduced coefficient amplitudes can restore noises, so that instance variables in original signals can be retained to the largest extent. However, such advantage of soft-threshold denoising is accompanied with worse denoising effect.

The formulas (3.1) and (3.2) respectively show hard-threshold denoising and soft-threshold denoising.

$$T\left(W_f,T\right)=\begin{cases} W_f & \left|W_f\right|\geq T \\ 0 & \left|W_f\right|<T \end{cases} \tag{3.1}$$

$$T\left(W_f,T\right)=\begin{cases} \text{sign}\left(W_f\right)\left(\left|W_f\right|-T\right) & \left|W_f\right|\geq T \\ 0 & \left|W_f\right|<T \end{cases} \tag{3.2}$$

The formula (3.3) is formed from (3.2):

$$T\left(W_f, T\right) = \begin{cases} \text{sign}\left(W_f\right)\left(1 - \dfrac{T}{|W_f|}\right)|W_f| & |W_f| \geq T \\ 0 & |W_f| < T \end{cases} \tag{3.3}$$

In the formula (3.3), the principle of soft-threshold denoising is that wavelet coefficients which exceed the threshold will converge to zero on the number axis in a certain proportion instead of being filtered directly. In most situations, soft-threshold denoising is an effective way to lower the risk of threshold selection and enhance robustness of wavelet threshold denoising.

3.2.3.2 Threshold Rules

The formulas (3.1), (3.2), and (3.3) prove that the selection of threshold value T is substantially important to the quality of denoising. There are four threshold selection rules: Sqtwolog, Rigrsure, Heursure, and Minimaxi.

Sqtwolog
The formula for threshold selection algorithm is:

$$T = \sigma\sqrt{2\ln N} \tag{3.4}$$

where J represents the scale of the wavelet transform, N represents the total number of wavelet coefficients on the scale of m, $(1 < m < J)$, after wavelet decomposition of the practical measurement signal $x(t)$, and σ represents the standard deviation of excess noise signals. Experiments prove that applying the rule of Sqtwolog can bring satisfactory denoising effect in the soft-threshold function.

Rigrsure
As a software threshold estimator, the rule of Rigrsure is a kind of adaptive threshold selection based on unbiased likelihood estimation. The likelihood estimate of a given threshold value T is to be found, firstly, and then the estimate will be a mini.

Let the signal $x(t)$ be a discrete-time sequence $(t = 1, 2, ..., N)$ and let $y(t)$ be an ascending sequence of $|x(t)|^2$.

The formula for calculating the threshold value is as follows:

$$\begin{cases} R(t) = \left(1 - \dfrac{t}{N}\right)y(t) + \dfrac{1}{N}\left(N - 2t + \sum_{i=1}^{t} y(i)\right) \\ T = \sqrt{\min\left(R(t)\right)} \end{cases} \tag{3.5}$$

Heursure
The rule of Heursure is a mixture of unbiased likelihood estimation and universal threshold principle that adopted at high SNR, and also at its low SNR adopted by the principle of fixed threshold.

The formula for calculating the threshold value is as follows:

$$
T = \begin{cases} T_1 & \dfrac{\|x(t)\|^2}{N} < 1 + \dfrac{1}{\sqrt{N}}\left(\log_2 N\right)^{3/2} \\[3ex] \min\left(T_1, T_2\right) & \dfrac{\|x(t)\|^2}{N} \geq 1 + \dfrac{1}{\sqrt{N}}\left(\log_2 N\right)^{3/2} \end{cases}
\tag{3.6}
$$

where N represents the length of the signal $x(t)$, T_1 represents the threshold value under the universal threshold principle, and T_2 represents the threshold value under the principle of unbiased likelihood estimation.

Minimaxi

The rule of Minimaxi is for fixed threshold values. In a given function set, the mean square error can be minimized to the largest extent.

The formula is as follows:

$$
T = \begin{cases} \sigma\left[0.3936 + 0.1829\ln(N-2)\right] & N \geq 32 \\[2ex] 0 & N < 32 \end{cases}
\tag{3.7}
$$

$$
\sigma = \frac{\text{Middle}\left(W_{1,k}\right)}{0.6745}, \quad 0 \leq k \leq 2^{j-1} - 1
\tag{3.8}
$$

In the formula, N represents the number of wavelet coefficients on the corresponding scale, $W_{1,k}$ represents wavelet coefficients on the scale of 1, j represents the scale of wavelet decomposition, and σ represents the standard deviation of noise signals.

Numerical Experiment

Let us take a rectangular wave signal whose SNR is 4. Wavelet threshold denoising is conducted for the signal under the four threshold rules mentioned above. Figure 3.1a and b respectively displays the original "Blocks" signal and the noisy "Blocks" signal. Figure 3.2a displays the effect of wavelet threshold denoising under the rule of Minimaxi, Figure 3.2b under the rule of Rigrsure, Figure 3.2c under the rule of Sqtwolog, and Figure 3.2d under the rule of Heursure.

Figure 3.2 shows that there is no big difference between the four denoising effects. In practice, which rule should be adopted depends on specific circumstance. A good threshold selection can contribute to retaining details of useful signals and eliminating noises as many as possible.

3.2.4   Online Implementation of Wavelet Denoising

Based on the threshold decision, the effect of wavelet filtering method compared with traditional filtering methods is relatively better. In practical projects, online wavelet threshold filtering is more significant and dependable than offline wavelet threshold filtering, which

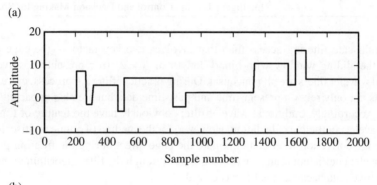

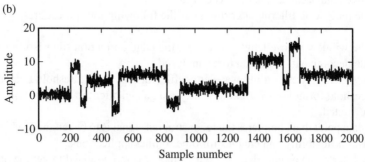

Figure 3.1 (a) The original "blocks" signal and (b) the noisy "blocks" signal.

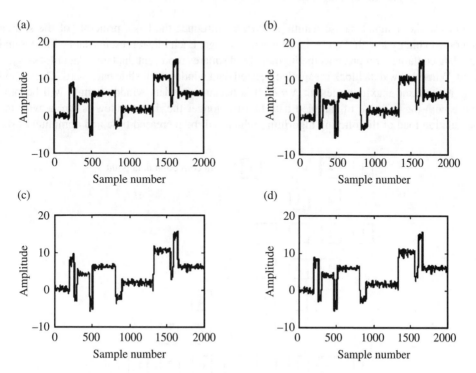

Figure 3.2 "Blocks" signal denoising under four threshold rules: (a) Minimaxi, (b) Rigrsure, (c) Sqtwolog, and (d) Heursure.

provide multiscale filtering access than that involves two key factors—the edge correction filter and the sliding window with binary length of $N = 2^n$ (n represents positive integer). Traditional wavelet filters are all noncausal. During practical filtering process, current wavelet coefficients not only rely on present-time and past-time data but are also impacted by future-time data. Accordingly, traditional wavelet filters obviously have the feature of time delay in terms of wavelet coefficient calculation. However, for the method of online multiscale filtering, a special edge correction filter can eliminate the edge effect during the filtering process and current wavelet coefficients can be worked out just through the filter algorithm without knowing future-time data because this filter is causal.

The online multiscale filtering algorithm has the following four procedures.

1. Within the window whose length is $N = 2^n$, the edge correction filter is used for wavelet multiscale decomposition of data to be analyzed.
2. After the signal decomposition, wavelet coefficients are to be threshold treated according to the threshold formula (3.7), and then the data gained after the threshold processing will be reconstructed.
3. The last data point that completes the reconstructed signal is to be retained, which improves the algorithm flexibility and enables for other online applications.
4. After the filter receives new data, the data window is to be moved to get the latest sample data. It should be ensured that the maximum window length is 2^k ($k=n$ or $k=n+1$). The window length will no longer increase when it reaches the maximum.

Let us take a signal whose length is $N=8$ to introduce the basic principle of the sliding window in binary length. Figure 3.3 shows the change. Each line represents data windows and the data contained; i represents the ith data; bold squares represent the latest data blocks sampled. When a new data block is sampled, related data window will slide backward. The sliding rule is the fixed maximum length $N=2^n$. It is because the data window length will become more with the increase of sampled data that the longer the data window is the heavier the calculation load of the filtering algorithm, which will be increased time and computer costs.

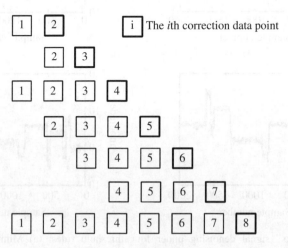

Figure 3.3 OLMS sliding window.

Therefore, the data window should remain unchanged when its length increases to a certain extent. In such a premise, the data window just slides for containing the latest data. Theoretically speaking, as long as the noise level at the current moment stays unchanged, the wavelet threshold value of each data window will be the same as to the data window length.

3.3 Data Consistency Analysis

Consistency analysis is an important means to correctness and reliability of testing data. Quantitative analysis of data consistency has been studied so far. In this section, the methods of statistical hypothesis testing and dynamic association analysis will be, firstly, introduced. Then the cluster analysis method will be focused based on the analysis of small sample characteristics and historical testing data.

3.3.1 Static Data Consistency Analysis

The consistency of testing methods, testing equipments, and testing environment is the foundation for testing data consistency. Static data can be classified as different types of subsamples according to the testing stage. Take a performance parameter of a certain unit device. The testing data of the parameter in the factory acceptance test and the testing data in the launch site test can be regarded as two different subsamples whose consistency will be analyzed afterward. The following content will show how to make consistency analysis of testing data which are obtained from two different testing stages.

For a certain parameter of a unit device, suppose the factory test data to be $X_1, X_2, \ldots, X_m$, and the launch site test data to be $Y_1, Y_2, \ldots, Y_n$. If the parameter under test follows a normal distribution, consistency testing of the two subsamples can be implemented by virtue of consistency testing of distribution parameters.

3.3.1.1 Consistency Testing of the Mean Value

If X_i and Y_i follow the distributions $N\left(\mu_1, \sigma_1^2\right)$ and $N\left(\mu_2, \sigma_2^2\right)$, respectively, the issue of mean value consistency then turns to the issue of following hypothesis testing:

$$H_0 : \mu_1 = \mu_2 \quad H_1 : \mu_1 \neq \mu_2 \tag{3.9}$$

The statistics is:

$$\mu = \frac{\left(\overline{X} - \overline{Y}\right) - \left(\mu_1 - \mu_2\right)}{\sqrt{\left(\sigma_1^2/m\right) + \left(\sigma_2^2/n\right)}} \tag{3.10}$$

We can get:

$$\mu \sim N\left(\frac{\left(\mu_1 - \mu_2\right)}{\sqrt{\left(\sigma_1^2/m\right) + \left(\sigma_2^2/n\right)}}\right) \tag{3.11}$$

If the hypothesis is valid, we can get:

$$\mu = \frac{\left(\bar{X} - \bar{Y}\right)}{\sqrt{\left(\sigma_1^2/m\right) + \left(\sigma_2^2/n\right)}} \tag{3.12}$$

The above formulas show the rule of consistency testing. After significance level α is given, the normal distribution quantile μ_α can be determined. If $|\mu| > \mu_\alpha$, it can be concluded that there is a big difference between X_i and Y_i. Otherwise, the mean values of X_i and Y_i are consistent. In practical projects, if σ_1 and σ_2 are unknown, unbiased estimation will be adopted as known variances:

$$\hat{\sigma}_1^2 = \frac{1}{m-1}\sum_{i=1}^{m}\left(X_i - \bar{X}\right)^2, \quad \hat{\sigma}_2^2 = \frac{1}{n-1}\sum_{i=1}^{m}\left(Y_i - \bar{Y}\right)^2 \tag{3.13}$$

3.3.1.2 Consistency Testing of Variances

If X_i and Y_i follow the distributions $N\left(\mu_1, \sigma_1^2\right)$ and $N\left(\mu_2, \sigma_2^2\right)$, respectively, the issue of variance consistency then turns to the issue of the following hypothesis testing:

$$H_0 : \sigma_1 = \sigma_2 \quad H_1 : \sigma_1 \neq \sigma_2 \tag{3.14}$$

By the fundamental theorem of statistics, we can get:

$$\frac{(m-1)\hat{\sigma}_1^2}{\sigma_1^2} \sim \chi^2(m-1), \quad \frac{(n-1)\hat{\sigma}_2^2}{\sigma_2^2} \sim \chi^2(n-1) \tag{3.15}$$

The statistics below follows the F distribution:

$$\frac{\hat{\sigma}_1^2/\sigma_1^2}{\hat{\sigma}_2^2/\sigma_2^2} \sim F(m-1, n-1) \tag{3.16}$$

If the hypothesis is valid

$$\frac{\hat{\sigma}_1^2}{\hat{\sigma}_2^2} \sim F(m-1, n-1) \tag{3.17}$$

The above formulas show the rule of consistency testing of variances. After the significance level is given, $F_{1-\alpha,2}(m-1, n-1)$ and $F_{\alpha,2}(m-1, n-1)$, the corresponding F distribution bilateral quantile, can be determined. If $F_{1-\alpha,2}(m-1, n-1) \leq \left(\hat{\sigma}_1^2/\hat{\sigma}_2^2\right) \leq F_{\alpha,2}(m-1, n-1)$, it can be concluded that the hypothesis is valid, and the variances of X_i and Y_i are consistent. Otherwise, there is a big difference between the variances of X_i and Y_i. The F distribution table only displays the upper quantile of some values of significance level α, so in practical cases $F_{1-\alpha}(f_1, f_2)$ can be worked out through the formula (3.18).

$$F_{1-\alpha}\left(f_1,f_2\right)=\frac{1}{F_\alpha\left(f_2,f_1\right)} \tag{3.18}$$

In practical use of the method of F testing, variances of two samples can be, firstly, worked out, and then let the bigger one be the numerator and the smaller one be the denominator. If so, F will always be greater than or equal to 1. It is only necessary to check the table for determining $F_{\alpha,2}(m-1,n-1)$. If $F>F_{\alpha,2}(m-1,n-1)$, it can be concluded that the hypothesis is invalid. Otherwise, the hypothesis can be accepted.

3.3.2   Dynamic Data Consistency Analysis

With the same input signals, multiple sets of output sequences are to be compared and analyzed in a pair for consistency testing of output data. If theoretical-output data sequence can be computed, such method can also be used to estimate the consistency of actual and theoretical output.

3.3.2.1   Statistical Hypothesis Testing

The statistical-hypothesis testing is a core component of statistical-inference and has a great significance for theoretical research and practical research applications. In mathematical-statistics, the statistical hypothesis is defined as an inference related to population distribution. Whether the hypothesis is valid or not depends on population samples. Here, the method is used to measure the consistency of two sets of the data. x_i and y_i are the two sets of data with equal length, and the differences of the two sets of data are only caused by self-measurements. $d_i = x_i - y_i$ is the difference value of x_i and y_i, and suppose that $d_1, d_2, ..., d_n$ are from the normal population $N(\mu_d, \sigma^2)$, which is unknown. If x_i is equal to y_i, difference values $d_1, d_2, ..., d_n$ are random errors. They are regarded to follow the normal distribution with zero mean. So, the hypothesis testing issue is:

$$H_0 : \mu_d = 0 \quad H_1 : \mu_d \neq 0 \tag{3.19}$$

Let a sample mean and variance of $d_1, d_2, ..., d_n$ be $\bar{d}$ and $\bar{\sigma}^2$ respectively. After the t testing of a single normal population mean, we can get the rejection region:

$$|t| = \left|\frac{\bar{d}-0}{\hat{\sigma}/\sqrt{n}}\right| > t_a\left(n-1\right) \tag{3.20}$$

If the testing result is not in the rejection region, it can be concluded that the null hypothesis is valid, which means there is no obvious difference between the two sets of data, so they are consistent. Usually in the t distribution table, n is less than 45. If the sample size is quite large, say over 50, it is shown by the central-limit theorem that when H_0 is valid, the statistics $U = (\bar{X} - \mu_d)/(s/\sqrt{n})$ will gradually follow $N(0, 1)$ and the rejection region is:

$$|U| = \left|\frac{\bar{X}-0}{\hat{\sigma}/\sqrt{n}}\right| > U_d \tag{3.21}$$

Hypothesis testing can be conducted for the statistics. Under normal circumstances, population variances are unknown, and it is necessary to use sample variances to estimate population variances. However, only when the sample size is over 100 can be the limit distribution that used to evaluate the approximate rejection region. In fact, when the sample size is quite large, the normal distribution and the t distribution are approximately equal to each other.

In many cases, whether the hypothesis testing is rejected largely depends on α, the significance test level. If the hypothesis is accepted, the greater the value α is the higher the similarity degree of the two sets of data and the higher the possibility that the hypothesis will be rejected. Thus, the data will be seen as an illegible, because the testing result may be inconsistent with the fact. However, if the value of α is too small, unqualified data may be recognized as qualified. Data consistency can be ranked according to the value of α. Usually-adopted values of α are $0.1, 0.05, 0.02, 0.01, 0.001$ which can be used as quantitative evaluation indexes of data consistency.

3.3.2.2 Dynamic Associated Analysis

The basic idea of the dynamic associated analysis is that both the sets of output data with the same input condition are viewed as a dynamic process—a time sequence— and then a scalar function on the two data sequence is constructed. The scalar function can be used as a qualitative index for evaluation of data consistency and dynamic associated. Specific procedures are as follows.

Let x_i and y_i be the two sets of output sequences and N be the data length. The scalar function below is defined as Theil's inequality coefficients (TIC):

$$\rho(x,y) = \frac{\sqrt{(1/N)\sum_{t=1}^{N}(x_t-y_t)^2}}{\sqrt{(1/N)\sum_{t=1}^{N}x_t^2}+\sqrt{(1/N)\sum_{t=1}^{N}y_t^2}} = \frac{\sqrt{\sum_{t=1}^{N}(x_t-y_t)^2}}{\sqrt{\sum_{t=1}^{N}x_t^2}+\sqrt{\sum_{t=1}^{N}y_t^2}} \tag{3.22}$$

Apparently, $\rho(x,y)$ has the following feature.

1. Symmetry: $\rho(x,y) = \rho(y,x)$
2. Normative: $0 \leq \rho(x,y) \leq 1$, $\rho = 0$ shows that the two sets of data sequences are completely consistent regardless of N, the data length. $\rho = 1$ shows that the two sets of data sequences are absolutely inconsistent.
3. The smaller ρ is, the better the consistency is. This is a non-statistic method and there is no limiting condition for the desired time sequence, so it is easy for application.

Statistical hypothesis testing and dynamic associated analysis mentioned in this section have already been engineering approaches, so they can be directly and widely applied in the area of data comparison in testing tasks. Analysis results got through these approaches can be seen as quantitative conclusions on data consistency, which has great significance to product performance assessment, fault removal, and quality control of key links.

3.3.3 Clustering Analysis Methods

Data are clustered according to their similarity level. After clustering, time-averaging operation is to be conducted for the same types of parameters. The mean curve serves as the basis for analysis of system operating conditions. That historical and current testing data clustered can help analyze the consistency of current and predict future testing data.

3.3.3.1 K-Mean Clustering Algorithm

1. Let x_l be the data to be clustered. $l = 1, 2, \ldots, N$. K values are then selected randomly as the initial clustering centers denoted by $Z_1(1)$, $Z_2(1), \ldots, Z_K(1)$. $Z_i(m)$, $i = 1, 2, \ldots, K$ stands for the ith clustering center after the mth iteration.
2. Data to be clustered in x_l are assigned to the above K clustering centers via the principle of minimum distance. If $\left\| x - Z_j(m) \right\| = \min \left\{ \left\| x - Z_i(m) \right\|, i = 1, 2, \ldots, K \right\}$ then $x \in C_j(m)$. m refers to the number of iterations; $C_j(m)$ refers to the jth cluster after the mth iteration; the clustering center is $Z_j(m)$.
3. A new clustering center is computed:

$$Z_j(m+1) = \frac{1}{N_j} \sum_{x_h \in C_j(m)} x_h; \quad j = 1, 2, \ldots, K; \ h = 1, 2, \ldots, N_j; \ N_j < N \tag{3.23}$$

N_j refers to the number of samples contained in the jth cluster $C_j(m)$. $x_h \subset x_l$.
4. If $Z_j(m+1) \neq Z_j(m)$, $j = 1, 2, \ldots, K$, let $m = m+1$. The steps 2 and 3 are to be repeated till $Z_j(m+1) = Z_j(m)$.

3.3.3.2 Clustering Analysis of Measurement Parameter Curves Based on K-Mean Clustering Algorithm

Measurement of Curve Similarity
Similarity of two curves can be reflected by similarity of their corresponding values and their change trend.

In Figure 3.4, for the curves L_a and L_b, similarity of their values can be defined as:

$$\text{Sim}_V(a,b) = \frac{1}{K} \sum_{k=0}^{T} \left[F_a(t_k) - F_b(t_k) \right]^2 \tag{3.24}$$

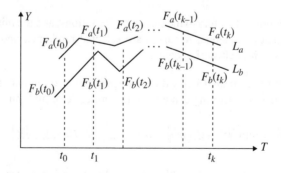

Figure 3.4 Similarity analysis on measurement parameter curves.

$F_a(t_k)$ and $F_b(t_k)$ are the values of L_a and L_b at the time of t_k. Similarity of their change trend can be defined as

$$\text{Sim}_T(a,b) = \frac{1}{K}\sum_{k=1}^{T}\left\{\left[F_a(t_k)-F_a(t_{k-1})\right]-\left[F_b(t_k)-F_b(t_{k-1})\right]\right\}^2 \tag{3.25}$$

Denotation of Similarity Degree of Parameter Curves

For historical data, parameter values of each sampling point are denoted as

$$F_n(1), F_n(2),..., F_n(t_k),..., F_n(T)$$

where $F_n(t_k)$ refers to the value at the time of t_k in the nth measurement and T refers to the end time of measurement. In the nth measurement and at the time of t_k, the changing value $Y_n(t_k)$ is

$$Y_n(t_k) = F_n(t_k)-F_n(t_{k-1}), \quad t_k = 1,...,T$$

Then we can get

$$Y_n(1), Y_n(2),..., Y_n(t_k),..., Y_n(T)$$

Sampling values of parameter curves and their change can be described as the formula below.

$$S_n = \left[F_n(0), F_n(1),..., F_n(T), Y_n(1), Y_n(2),..., Y_n(T)\right] \tag{3.26}$$

where n is a certain parameter in the nth launch. $n = 1, 2,..., N$, and N refers to the number of launches.

Steps of the Algorithm

Step 1. Outliers elimination

Step 2. Selecting K initial clustering centers

$$Z_1(1) = [S_{a1}], \quad Z_2(1) = [S_{a2}], \quad Z_K(1) = [S_{aK}]$$

The number in the brackets refers to the number of iterations, and the initial number is 1. Of historical measurement parameters, K curves are selected randomly, and the corresponding $S_{a1}, S_{a2},..., S_{aK}$ are just initial clustering centers.

Step 3. According to the formula (3.24), the corresponding S_n of all historical curves will be assigned to the above K clustering centers via the principle of minimum distance.

$$\left\|S_n - Z_j(m)\right\|_2 = \min\left\{\left\|S_n - Z_i(m)\right\|_2, \ n = 1, 2,..., N, \ i = 1, 2,..., K\right\}, \ S_n \in C_j(m)$$

$C_j(m)$ refers to the jth cluster after the mth iteration. Its clustering center is $Z_j(m)$. N refers to the number of measurement.

Step 4. A new clustering center is computed:

$$Z_j(m+1) = \frac{1}{N_j} \sum_{S_i \in C_j(m)} S_i, \quad j = 1, 2, \ldots, K \tag{3.27}$$

N_j refers to the number of curves contained in the jth cluster $C_j(k)$. The formula (3.27) represents that the values of the jth clustering center are mean values of S_n in this clustering center.

By the formula (3.26) we can get:

$$S_i = \left[F_i(0), F_i(1), \ldots, F_i(T), Y_i(1), Y_i(2), \ldots, Y_i(T) \right]$$

Then the formula (3.27) can turn into the formula (3.28).

$$
\begin{aligned}
Z_j(m+1) = \frac{1}{N_j} \Bigg\{ &\sum_{S_{l1},\ldots,S_{Nj} \in C_j(m)} \left[F_{l1}(0) + F_{l2}(0) + \cdots + F_{lN_j}(0) \right], \ldots, \\
&\sum_{S_{l1},\ldots,S_{Nj} \in C_j(m)} \left[F_{l1}(T) + F_{l2}(T) + \cdots + F_{lN_j}(T) \right], \\
&\sum_{S_{l1},\ldots,S_{Nj} \in C_j(m)} \left[Y_{l1}(1) + Y_{l2}(1) + \cdots + Y_{lN_j}(1) \right], \ldots, \\
&\sum_{S_{l1},\ldots,S_{Nj} \in C_j(m)} \left[Y_{l1}(T) + Y_{l2}(T) + \cdots + Y_{lN_j}(T) \right] \Bigg\}
\end{aligned}
\tag{3.28}
$$

N_j refers to the number of samples contained in the jth cluster $C_j(m)$.

Step 5. If $Z_j(m+1) \neq Z_j(m)$, $j = 1, 2, \ldots, K$, let $m = m+1$. The steps 3 and 4 are to be repeated till $Z_j(m+1) = Z_j(m)$. Finally, we get the cluster $C_j(m+1)$, $j = 1, 2, \ldots, K$.

The corresponding curve of the jth clustering center $(j = 1, 2, \ldots, K)$ is:

$$
\begin{aligned}
z_j(t_k) = \frac{1}{N_j} \Bigg\{ &\sum_{S_{l1},\ldots,S_{Nj} \in C_j(m)} \left[F_{l1}(0) + F_{l2}(0) + \cdots + F_{lN_j}(0) \right], \ldots, \\
&\sum_{S_{l1},\ldots,S_{Nj} \in C_j(m)} \left[F_{l1}(T) + F_{l2}(T) + \cdots + F_{lN_j}(T) \right] \Bigg\}
\end{aligned}
$$

where $t_k = 1, \ldots, T$.

After clustering analysis of a certain parameter curve in Figure 3.5, three clustering center curves are computed, as shown in Figure 3.6.

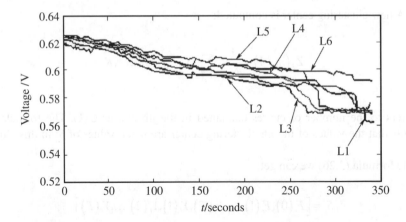

Figure 3.5   Changing curve of a certain parameter.

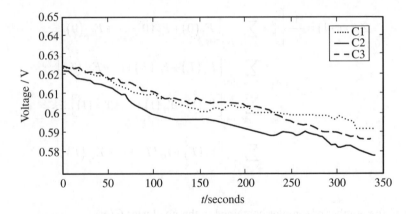

Figure 3.6   Clustering results for normal data.

3.3.3.3   Dynamic Prediction on Parameters Based on Clustering Analysis

After clustering analysis of historical data, the trend of normal values of a certain parameter can be computed. As Figure 3.6 shows, the three different trends are similar to each other. Classification of normal historical parameter data based on clustering analysis can contribute to prediction on the parameter in the system.

$F_{new}(t_i)$ represents current data of a given parameter. The measured parameter before the current time of t_k is $s_{new}(t_{k-1}) = [F_{new}(0), F_{new}(1), ..., F_{new}(t_{k-1})]$. According to the formula (3.24), we can calculate the similarity between the system status and the corresponding clustering center curve.

$$d(i, j) = \left\| s_{new}(i) - z_j(i) \right\|_2$$

In the formula, $z_j(i)$ represents the value of the jth clustering center curve before the time of i.

$$z_j(i) = \frac{1}{N_j} \left\{ \sum_{S_{l1},\ldots,S_{Nj} \in C_j(m)} \left[F_{l1}(0) + F_{l2}(0) + \cdots + F_{lN_j}(0) \right], \ldots, \right.$$

$$\left. \sum_{S_{l1},\ldots,S_{Nj} \in C_j(m)} \left[F_{l1}(i) + F_{l2}(i) + \cdots + F_{lN_j}(i) \right] \right\}$$

where $d(i, j)$ represents the distance between $s_{new}(i)$ and $z_j(i)$ at the time of i.

$T(j)$ refers to the total number of points on the current measured curve with the nearest distance to the jth cluster S_j during the time of $[0, t_k)$ ($j = 1,2,\ldots,K$). The calculation rules are as follows.

1. When $i = 1,2,\ldots,K$, calculate $T(j)$, the total number of points in the current measured data $F_{new}(t_i)$ $\{F_{new}(t_i) = \min[d(i, 1), d(i, 2), \ldots, d(i, K)]\}$ with the nearest distance to the jth clustering center curve at the time of t_i ($j = 1, 2, \ldots, K$).
2. Get the maximum $T(j)$. After ranking $T(1), T(2), \ldots, T(K)$, work out the value of j for the maximum $T(j)$.
3. If several values of j are computed, change the time period $[0, t_k)$ into $[1, t_k)$, and repeat the steps 1 and 2 until there is just only one value of j.

$T(j)$ can be computed according to the above steps ($j = 1, 2, \ldots, K$). The value on the corresponding curve of the jth clustering center at the time of t_{k+1} (j here is for the maximum $T(j)$) is taken as a reference value for this parameter at the time of t_{k+1}.

The clustering curves can reveal the relation between parameter change and time when a system is working in a normal state that can be used to predict the working state of systems. By the above algorithm, we can get the clustering curve closest to measured values. For a parameter, if $d_{min}(t_k, j) < \delta$ (δ is the allowable variation range of the parameter), it can be concluded that the system is working in a normal state at the current time of t_k and will work in a normal state at the next moment of t_{k+1}. While if $d_{min}(t_k, j) < \delta$ at the time of t_k, which means the parameter deviates from the clustering curve, it can be concluded that the system is being abnormal; if $d_{min}(t_{k+1}, j) > \delta, d_{min}(t_{k+2}, j) > \delta, \ldots, d_{min}(t_{k+n}, j) > \delta$, it can be predicted that the system will be abnormal after the time of t_k.

Figure 3.7 shows a comparison between a parameter's actual data and its historical trend. The figure indicates that the change of real parameter values is consistent with the clustering trend of historical data. Within the Specified range, the system is regarded as normal.

Figure 3.8 shows a comparison between a parameter's actual data and its clustering-trend when the system is abnormal. Before 330 seconds, the trend of real data is consistent with that of the clustering curve, which means the system is normal at this period of time. However, after 330 seconds, the deviation between actual data and the clustering trend is widening, which reveals that the parameter begins to be abnormal, so does the system.

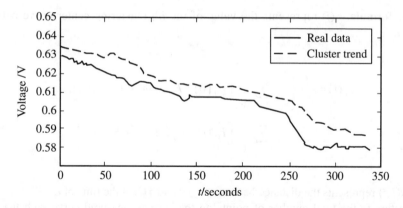

Figure 3.7 Comparison between normal data and clustering trend.

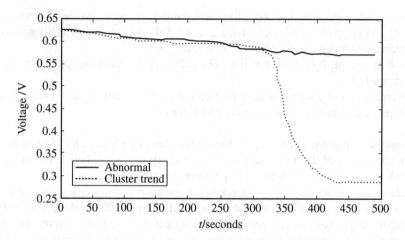

Figure 3.8 Comparison between abnormal data and clustering trend.

3.4 Analysis of Data Singular Points

Singular signals refer to transient signal in testing data, which is important data features that contain much information about the testing subject. Transient signal points often contain essential information about the rocket launch process. Detecting accurate catastrophe point are very significant for rocket separation and security control.

Detecting and analyzing a singular point's process is not an easy task, which faced various kinds of abrupt signals noises. Traditional methods based on Fourier transform cannot make a topical analysis and detect in the time domain and transient signal. On the other hand, wavelet transform can make a local analysis of signals and adjust the width of the time-frequency window as required. Therefore, wavelet transform is regarded as an effective method for detection and analysis of transient signal.

3.4.1   Signal Singularity

In mathematics, singularity of the signal $f(x)$ is described by Lipschitz index.

Let n be nonnegative integers and $n < \alpha \leq n+1$. If there exist two constants M and $h_0 (M > 0, h_0 > 0)$ and the polynomial $g_n(h)$, when $h < h_0$ and

$$\left| f(x_0 + h) - g_n(h) \right| \leq M |h|^\alpha \tag{3.29}$$

Then $f(x)$ at the point x_0 is the type of Lipschitz-α. If for all x_0, $x_0 + h \in (a, b)$ and the formula (3.29) is valid, then it can be said that $f(x)$ in the interval (a, b) is the consistent Lipschitz-α cluster.

Let $f(x)$ be continuous signals. If $f(x)$ at the point x_0 is not the type of Lipschitz-1, $f(x)$ is seen as a singular at the point x_0. Some conclusions are drawn on signal singularity as follows.

1. The bigger the Lipschitz index of $f(x)$ is, the smoother $f(x)$ will be.
2. If the function is continuous, differentiable, or discontinuous at one certain point but the derivative is bounded, its Lipschitz index is always 1.
3. If the Lipschitz index of $f(x)$ at the point x_0 is less than 1, $f(x)$ is seen as singular at this point.

In summary, the Lipschitz index of $f(x)$ at the point x_0 reflects the singularity of $f(x)$ at this point.

The Lipschitz index can be evaluated through its definition, but it is quite complicated and the noise effect has not been considered yet. Wavelet transform has the good feature of localization, and it can well reflect topical singularity of signals. For any isolated singular point, it can be known that wavelet transform will get the sectional modulus maximum value if the approaching-to-zero speed of the absolute value of the wavelet coefficient is slower than that of the absolute value of its neighborhood wavelet coefficient. Thus, wavelet transform can locate singular points and make quantitative description of local singularity of signals.

Some conclusions are drawn on determining singularity of $f(x)$ at the point x_0 by wavelet transform.

Suppose that the wavelet basis has n vanishing moments and they are differentiable with compact support. n represents positive integers and $\alpha \leq n$, $f(x) \in L^2(R)$. In the neighborhood of x_0 and at other scales, if a constant A follows the formula (3.30), then the Lipschitz index of $f(x)$ at the point x_0 is α.

$$\left| Wf(s, x) \right| \leq A \left(s^\alpha + |x - x_0|^\alpha \right) \tag{3.30}$$

The formula above reflects the relation between wavelet transform and the Lipschitz index of $f(x)$ at the point x_0. The formula (3.30) shows that singular points are distributed on the modulus line. The Lipschitz index is not equal to 1 and $\alpha > 0$, so the signal is singular. Therefore, wavelet transform can be utilized to detect classification.

Let x_0 be a local singular point of $f(x)$. At this point, wavelet transform of $f(x)$ has the modulus maximum. When it comes to discrete dyadic wavelet transform, the formula (3.30) turns into the formula (3.31).

$$\left| W_{2^j} f(s, x) \right| \leq K \left(2^j \right)^\alpha \left(1 + |x - x_0|^\alpha \right) \tag{3.31}$$

where j is the binary scale parameter and x is the discrete value. The formula (3.32) can be computed from the formula (3.31).

$$\log_2\left|W_{2^j}f\left(x\right)\right|\leq\log_2 K+\alpha j+\log_2\left(1+\left|x-x_0\right|^{\alpha}\right) \tag{3.32}$$

If the singularity index of $f(x)$ at the point x_0 is greater than zero, it can be known from the formula (3.31) that with the increase in j, the logarithm of the modulus maximum of wavelet transform will also increase.

3.4.2 Determination of Signal Singular Point Location

There are smooth functions $\theta(x)$, $\theta(x)=O[1/(1+x^2)]$, and $\int_R\theta(x)dx\neq 0$. Besides, $\theta_s(x)=1/[s\theta(x/s)]$. The formula (3.33) contains two wavelet transform functions.

$$\psi^1\left(x\right)=\frac{d\theta\left(x\right)}{dx},\psi^2\left(x\right)=\frac{d^2\theta\left(x\right)}{dx^2} \tag{3.33}$$

If $f(x)\in L^2(R)$, its wavelet transform can be

$$W^1f\left(s,x\right)=f*\psi_s^1\left(x\right)=s\frac{d}{dx}\left(f*\theta_s\right)\left(x\right) \tag{3.34}$$

$$W^2f\left(s,x\right)=f*\psi_s^2\left(x\right)=s^2\frac{d^2}{dx^2}\left(f*\theta_s\right)\left(x\right) \tag{3.35}$$

$f*\theta(x)$ can make $f(x)$ smooth. For any scale s, $W^1f(s,x)$ and $W^2f(s,x)$ are respectively in direct proportion to the first-order derivative and the second-order derivative of $f*\theta(x)$.

Figure 3.9 shows that singular points on $f(x)$, through wavelet transform, are reflected as maximum values on $W^1f(s,x)$ while as zero crossing points on $W^2f(s,x)$. Therefore, to determine the location of singular points is just to evaluate the optimal values on $W^1f(s,x)$ or the zero crossing points on $W^2f(s,x)$. It is better to evaluate the maximum values on $W^1f(s,x)$.

The maximum values on $W^1f(s,x)$ have the feature of transmission with the change of s. Mallat used to prove that if the wavelet at a smaller scale has no local modulus maximum,

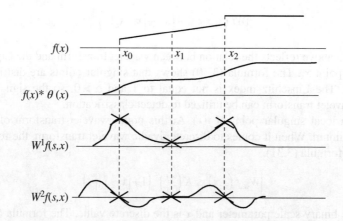

Figure 3.9   Relational figure for $f(x)$, $f*\theta(x)$, $W^1f(s,x)$, and $W^2f(s,x)$.

there cannot be any singular points in the neighborhood. This shows that the existence of singular points has close relation to the modulus maximum at each scale. Normally, modulus maximum points forming a modulus maximum curve are just seen as singular points.

3.4.3 Numerical Examples

3.4.3.1 Example 1

There is a curve in a certain testing process, shown in Figure 3.10a. Five layers of wavelet decomposition have been conducted to a pulse signal. Figure 3.10b reflects the detail signals of each layer. It can be found in Figure 3.10b that the location of singular points can be accurately

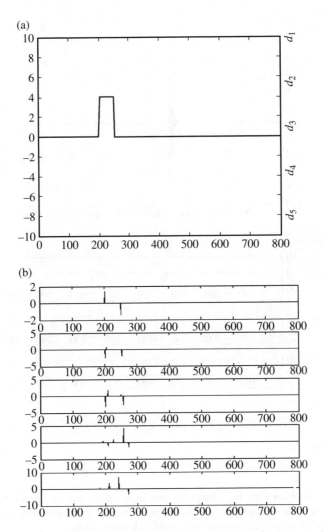

Figure 3.10 An transient signal and its detail signal waveform after wavelet decomposition. (a) Transient signal and (b) detail high-frequency wave after wavelet decomposition.

determined in detail signals d_1 and d_2 but not in d_3, d_4, or d_5. Therefore, it can be concluded that by wavelet transform the location of singular points of pulse signals can be detected accurately in d_1 and d_2.

3.4.3.2 Example 2

There is a signal carrying high-frequency messages locally, shown in Figure 3.11a. Figure 3.11b reflects the high frequency message of each layer after five layers of wavelet decomposition of this signal. In Figure 3.11b, it can be found that singular points on the first and the second layers can be located accurately, but not on the third, fourth, or fifth layers.

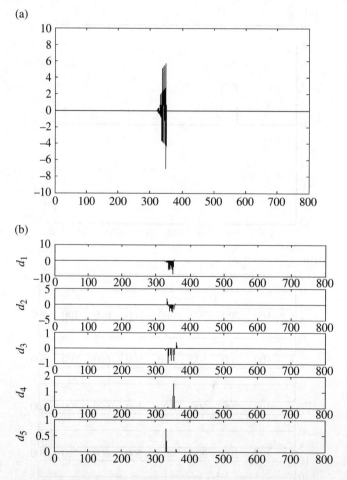

Figure 3.11 A signal carrying high-frequency message locally and its layers of high-frequency message after wavelet decomposition. (a) Waveform of the signal carrying high-frequency message locally and (b) detail high-frequency wave after wavelet decomposition.

3.4.3.3 Example 3

There is a slowly changing signal shown in Figure 3.12a. Figure 3.12b reflects the detail signals of each layer after five layers of wavelet decomposition of this slowly changing signal. In Figure 3.12b, singular points near the point 650 can be displayed clearly in d_1 and d_2. However, the singular points near the point 550, not the point 650, are displayed clearly in d_5. This shows that it has some limitations to detect singular points of slowly changing signals by wavelet decomposition.

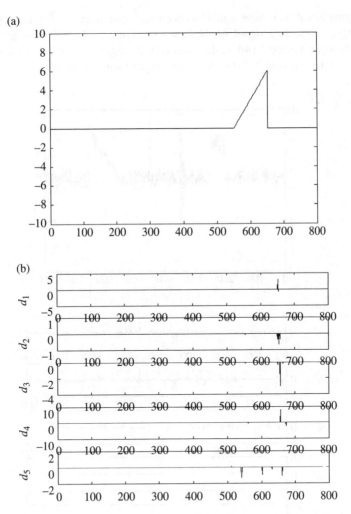

Figure 3.12 A slowly changing signal and its layers of high-frequency message after wavelet decomposition. (a) Waveform of the slowly changing signal and (b) detail high-frequency message after wavelet decomposition.

3.4.3.4   Example 4

There is a mixed signal carrying Gaussian noises shown in Figure 3.13a. Figure 3.13b reflects the detail signals of each layer after five layers of wavelet decomposition of this mixed-signal. In Figure 3.13b, it can be found that with noises, singular points of the original signal cannot be displayed in d_1 and d_2, and singular points of the slowly changing signal cannot be displayed clearly either in d_4 or d_5. This shows that detecting singular signals by wavelet decomposition will be largely effected by noises.

3.4.4   Experimental Examples

A given engine vibration testing signal has been analyzed in terms of singular points and the rate of change. The testing signal A6 is shown in Figure 3.14a and its singularity analysis results are shown in Figure 3.14b, c, d, e, and f. In the figure, "*" stands for a singular point and the line segment linking "*" stands for the range of singular points.

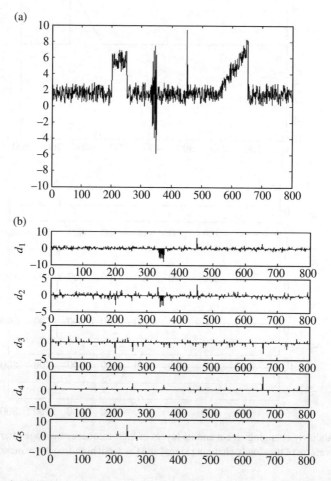

Figure 3.13   A mixed signal carrying Gaussian noises and its detail signals of each layer after five layers of wavelet decomposition. (a) A mixed signal carrying Gaussian noises and (b) detail high-frequency message after wavelet decomposition.

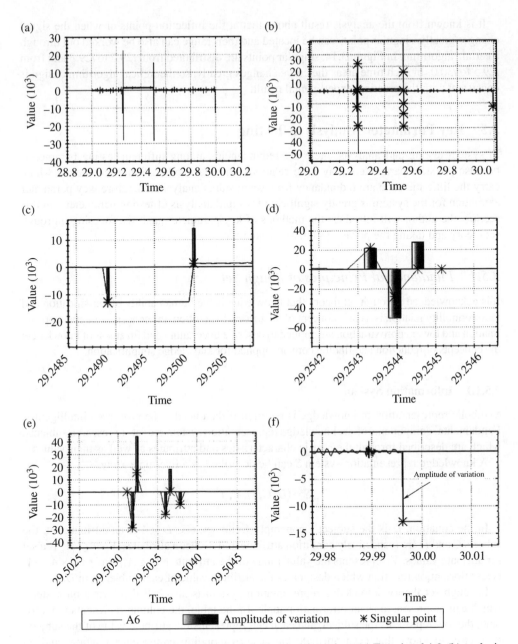

Figure 3.14 Analysis result of test signal A6 and its singular points. (a) Test signal A6, (b) analysis result of singular points of A6, (c) local amplification of singular points from 29.2485s to 29.2505s, (d) local amplification of singular points from 29.2542s to 29.2546s, (e) local amplification of singular points from 29.5025s to 29.5045s, and (f) local amplification of singular points from 29.97s to 30.01s.

It is known from the analysis result above that at the inflection points or when the signal changes greatly, singular points can be located and their range can also be worked out through a certain algorithm. In Figure 3.14, singular points are distributed from $t29.2$ to $t29.3$ and from $t29.5$ to $t30.0$. In such time areas the curve changes greatly, so theoretical algorithm design is consistent with the practical experimental result.

3.5   Key Parameter Analysis in Testing

In system testing and analysis, there are redundancies among different testing parameters because physical parameters may have relations with each other. Some parameters which carry the little message are redundancy for system status analysis. Therefore, key parameter extraction for the system is greatly significant for rapid analysis of testing parameters. In this section, the author will introduce the methods of key parameter analysis, focusing on rough set theory and taking the power system as an example.

3.5.1   Fundamental Principles of Rough Set

It is acknowledged in rough set theory that knowledge is the human ability of classification and indiscernibility relation is a basic concept of the theory. In addition, the concepts of upper approximation and lower approximation are applied to describe uncertainty and fuzziness of knowledge; the concepts of reduction and finding core are applied for knowledge simplification.

3.5.1.1   Information System

Symbolic representation of knowledge is required if data need to be processed intelligently. Fundamental components of the knowledge representation system are sets of research subjects, which are described through their basic characteristics and attributes as well as their values.

A knowledge representation system S can be described as a five-tuple array.

$$S = \langle U, C, D, V, f \rangle$$

In the equation, U is the subject's nonempty finite set, $C \cup D = A$ is an attribute set, the subsets C and D are respectively condition attribute and decision attributes, $V = \cup_{a \in A} V_a$ is a set for attribute values, V_a represents the value range of the attribute a ($a \in A$), and $f : U \times A \to V$ is an information function which designates the attribute value of each subject x in U.

In rough set theory, knowledge representation system is also called information system, which can be presented as an information table. In the table, the column represents the attribute, the row represents the subject, and each cell represents the attribute value of the subject. It is easy to know that each attribute has a corresponding equivalence relation, and an information table can be seen as a set of equivalence relations, namely a knowledge base.

3.5.1.2   Rough Set

A set is defined by elements in it. The set is fixed if each element in it is uniquely fixed. The definition of a set is explicit in mathematics or it cannot prove any mathematical theorem.

As a continuation of classical set theory, rough set theory introduces the knowledge that is for classification as a part of a set. Whether the subject α belongs to the set X depends on the knowledge of domain. There are three situations.

1. α definitely belongs to X.
2. α definitely does not belong to X.
3. α may belong to X and may not.

Set partitioning depends on the knowledge of relative than an absolute domain. U is a given domain, and equivalence relation R is partitions of U into different basic equivalence classes U/R, which are mutually disjoint. Let X be a set in U. If X represents a union set made up of equivalence classes, X is definable by R in U or else X is indefinable by R. Being definable by R refers to be defined exactly in the knowledge base $K = (U, R)$. Whereas, indefinable by R refers to be impossible to be defined in this knowledge base. R-definable sets are also called R-exact sets. Similarly, R-indefinable sets are called R-inexact sets or rough sets.

If $R \in \mathrm{IND}(K)$ and X is a R-exact set, then the set X ($X \subseteq U$) is an exact set in K; if $R \in \mathrm{IND}(K)$ but X is a R-inexact set or a rough set, then X is a rough set in K. A rough set can be defined approximately, which is achieved through approximating of two exact sets—upper and lower approximating rough set.

There is a knowledge base $K = (U, R)$ and the classification U/R. For each subset X ($X \subseteq U$), the following two sets are respectively R lower approximation and R upper approximation of X.

$$R_-(X) = \left\{ x \in U : [X]_R \subseteq X \right\}$$
$$R^-(X) = \left\{ x \in U : [X]_R \cap X \neq \Phi \right\}$$

If and only if $[X]_R \subseteq X, x \in R_-(X)$; subject to: $[X]_R \cap X \neq \Phi, x \in R^-(X)$

$R_-(X)$ is a set that contains the elements in the knowledge R, U that definitely belong to X; $R^-(X)$ is a set that contains the elements in the knowledge R, U that may belong to X.

The above is the definitions of R lower approximation and R upper approximation. The next part is about definitions of positive domain, negative domain, and boundary domain.

3.5.1.3 Positive Domain, Negative Domain, and Boundary Domain

In Figure 3.15, $\mathrm{POS}_R(X) = R_-(X)$ is R positive domain of X; $\mathrm{NEG}_R(X) = U - R^-(X)$ is R negative domain of X; $\mathrm{BN}_R(X) = R^-(X) - R_-(X)$ is the boundary domain of X.

$\mathrm{POS}_R(X)$ or lower approximation of X is a set that contains the elements in the knowledge R that definitely belong to X. Similarly, $\mathrm{NEG}_R(X)$ is a set that contains the elements in the knowledge R that definitely do not belong to X, namely a complementary set of X. Boundary domain is a uncertain domain to some extent. In the knowledge R, it is not certain whether elements in the boundary domain belong to X or $-X$. Upper approximation of X is a set that contains the elements in the knowledge R that may or may not belong to X. In form, upper approximation is a union set of positive domain and boundary domain.

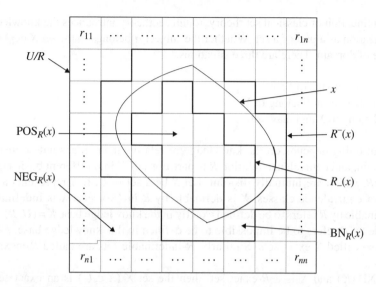

Figure 3.15   Positive domain, negative domain, and boundary domain.

3.5.1.4   The Importance of Classification Feature and Attribute of Rough Set

Let U be the discourse domain for data description in rough set theory. Let X be a subset in the domain, so $X \subseteq U$. Let R be equivalence relation for attribute representation of system parameters. If X can be described precisely by R attribute set, X is seen as R-definable. But in reality, approximation classification may be uncertain, which means some elements in U definitely belong to X while some may not belong to X.

The following two subsets are taken into consideration in order to measure whether the description of the basic set Y based on R is proper for the uncertainty of approximation classification of X.

$$R_-\left(X\right) = \cup\left\{Y \subseteq U \mid R : Y \subseteq X\right\}$$
$$R^-\left(X\right) = \cup\left\{Y \subseteq U \mid R : Y \cap X \neq \Phi\right\}$$

The above two sets are respectively R lower approximation and R upper approximation. $R_-(X)$ is a set that contains the elements in the attributes R and U that definitely belong to X. $R^-(X)$ is a set that contains the elements in the attributes R and U that definitely or may belong to X. $\mathrm{BN}_R(X)$ is R boundary for X, a set that contains the elements in the attributes R and U that do not belong to X definitely or to $-X$ definitely.

It is certain that the subjects in U belong to X or $-X$, two uncorrelated subsets. The total number of subjects in U equals to the number of subjects that belong to R rather than X.

$$\left|U - \mathrm{BN}_R\left(X\right)\right| = \left|U\right| - \left|R^-\left(X\right) - R_-\left(X\right)\right|$$

where $|\cdot|$ represents cardinal number or cardinality of a set. As for a finite set, $|\cdot|$ represents the number of elements contained in the set.

If R is only a given parameter of the system, parameter importance of the system can base on its classification ability.

$$\alpha_R(X) = \frac{|U| - |R^-(X) - R_-(X)|}{|U|} \qquad (3.36)$$

The following formulas can explain the classification in the formula (3.36); see Figure 3.15 for a reference. Let $n = 8$, and U/R represents equivalence relation R in the discourse domain U. Let three subsets R_1, R_2, R_3 are as follows.

$$R1 = \left(r_{11}, \ldots, r_{18}, r_{21}, r_{22}, r_{26}, r_{27}, r_{28}, r_{31}, r_{37}, r_{38}, r_{41}, r_{48}, r_{51}, r_{58}, r_{61}, r_{62}, r_{68}, r_{71}, r_{72}, \right.$$
$$\left. r_{73}, r_{77}, r_{78}, r_{81}, \ldots, r_{88} \right)$$
$$R2 = \left(r_{34}, r_{43}, r_{44}, r_{45}, r_{54}, r_{55}, r_{56}, r_{65} \right)$$
$$R3 = \left(r_{23}, r_{24}, r_{25}, r_{32}, r_{33}, r_{35}, r_{36}, r_{42}, r_{46}, r_{47}, r_{52}, r_{53}, r_{57}, r_{63}, r_{64}, r_{66}, r_{67}, r_{74}, r_{75}, r_{76} \right)$$

Elements contained in $|U| - |R^-(X) - R_-(X)|$ belong to either R_1 or R_2. Elements contained in $BN_R(X) = R^-(X) - R_-(X)$ belong to R_3. R_1 does not belong to X for sure, while R_2 does belong to X. According to the classification of elements in U by R, it is certain that elements in R_1 do not belong to X, and elements in R_2 belong to X. It can depend on R to make an understandable classification of elements in R_1 and R_2, which are closely related to X, or in other words, have a clear impact on X. However, it cannot depend on R to decide whether elements in R_3 belong to X, so the relevance between R_3 and X cannot be determined either. $\left(|U| - |R^-(X) - R_-(X)| \right)$, the numerator in the formula (3.36), represents the elements in R_1 that does not belong to X and the elements in R_2 that belongs to X. $|U|$, the denominator in the formula (3.36), represents the total number of elements in the domain U. Therefore, the formula (3.36) represents the proportion of elements, which have comprehensible relation with X in the domain U. The formula (3.36) can be regarded as a basis for judging the correlation between R and X.

In conclusion, $\alpha_R(X)$ shows the method of using R to describe the membership of subjects in X and represents the effectiveness of classification by the feature or feature set, which can be used to measure the classification ability of a certain feature R. The greater $\alpha_R(X)$ is, the stronger the classification ability of R can be proved to be.

3.5.2   Simplifying Parameters by Rough Set

3.5.2.1   Discretization of Parameter Values Based on Self-Organizing Feature Maps

It is a must to discretize the testing data related to the power system, if rough set theory is applied for parameter data analysis. Discretization of continuous attribute values refers to partitioning condition attribute values into several discrete subintervals in which discrete values replace original actual values of the parameter. Such process can be seen as the issue of clustering analysis. Hence, continuous attribute values can be discretized by the method of clustering.

Because of the indeterminate distance between parameter values, a self-organizing feature map, also called a Kohonen network, is a better way for parameter data clustering.

First described by the Finnish Scholar Teuvo Kohonen as an artificial neural network, a self-organizing feature map is a feed forward neural network in the mode of unsupervised competitive learning. Self-organizing maps are different from other artificial neural networks in the sense that they use a neighborhood function to preserve the topological properties of the input space. A self-organizing map consists of components called nodes or neurons. The link weight of each neuron is distributed in a certain way and adjacent neurons excite each other mutually. The basic principle of the Kohonen network is elaborated as follows.

In a certain input pattern, a node of the output layer is excited to the largest extent and becomes a winner node. At the same time, some nodes in its neighborhood are also excited due to lateral action. Then the weight vectors connected with these nodes will be modified toward the direction of input pattern. If the input pattern changes, the winner node will also change from its original form. In such way, the network can adjust the weight in a self-organizing way through a large number of training sample data. Finally, the network output layer can reflect the general distribution of the whole data. In other words, the basic data distribution features can be shown in sample data.

Procedures of discretization through Kohonen networks are as follows.

1. Structural parameters of neurons should be determined. The number of neurons in the competitive layer is the final classification number, the learning rate of weights, the learning rate of thresholds, and the total number of learning steps.
2. Each parameter's value is taken as an input sample $I = (x_0, x_1, x_2, ..., x_K)$.
3. After the learning of data distribution features via Kohonen networks, M clustering centers are computed and ranked in an ascending order $C_1, C_2, ..., C_M$.
4. The middle point of two adjacent clustering centers is taken as the boundary of discretization.

$$B_n = \frac{C_n + C_{n+1}}{2}$$

In the formula, $n = 1, 2, 3, ..., M - 1$. Discretization intervals are $(-\infty, B_1], (B_1, B_2], (B_2, B_3], ...,$ $(B_{M-2}, B_{M-1}], (B_{M-1}, +\infty)$. After discretization for each interval, corresponding discrete values can be computed.

For example, in the power system, some parameters are redundancies for the reflection of the system's condition. These parameters carry very little information. Analyzing the importance of parameters related to pressure condition can help get key parameters that influence pressure variation of the storage tank.

According to the structure principle of the power system, the key parameters include pressure of blow-off cylinder (parameter a), outlet pressure of cold helium pressure regulator (parameter b), outlet flow of pressure regulator (parameter c), inlet pressure of oxygen pump (parameter d), pressure of thrust chamber (parameter e), outlet pressure of pressure regulator (parameter f), outlet pressure of liquid oxygen (parameter g), liquid oxygen level (parameter h), and pressure of oxygen tank (parameter i). These parameters are discretized via self-organizing feature maps and then the discretized parameter values are obtained, as shown in Table 3.1.

Table 3.2 is an attribute parameter table obtained after getting rid of the duplicate values in Table 3.1.

Table 3.1 Attribute values after discretization.

a	b	c	d	e	f	g	h	i
5	2	1	1	1	5	5	5	5
5	2	1	1	1	5	5	5	5
5	2	1	1	1	5	5	5	5
4	2	1	1	1	5	5	5	5
4	2	5	1	1	5	5	5	5
3	2	5	1	1	5	5	5	5
3	5	5	1	1	5	5	5	5
3	5	5	5	1	5	5	5	5
3	5	5	5	1	5	5	5	5
3	5	5	5	5	5	5	5	4
1	1	5	1	1	1	2	1	1
1	1	5	1	1	1	1	1	1
1	1	5	1	1	1	1	1	1
1	1	1	1	1	1	1	1	1
1	1	1	1	1	1	1	1	1
...	...	...	...	...	...	...	...	...

Table 3.2 Attribute parameters after getting rid of duplicate values.

Sample	a	b	c	d	e	f	g	h	i	Status of parameter i
X_1	1	1	1	1	1	1	1	1	2	W_2
X_2	1	1	5	1	1	1	1	1	1	W_1
X_3	1	2	5	1	1	1	1	1	1	W_1
X_4	1	5	5	1	1	1	1	1	1	W_1
X_5	1	5	5	2	1	1	1	1	1	W_1
X_6	1	5	5	2	2	1	2	1	1	W_1
X_7	2	5	5	2	5	2	5	1	1	W_1
X_8	3	2	5	5	1	5	1	5	5	W_5
X_9	3	5	5	1	5	2	5	5	1	W_1
X_{10}	3	5	5	1	5	3	5	5	1	W_1
X_{11}	3	5	5	1	5	4	5	5	1	W_1
X_{12}	3	5	5	1	5	5	5	5	1	W_1
X_{13}	3	5	5	2	5	2	5	1	1	W_1
X_{14}	3	5	5	2	5	2	5	5	1	W_1
X_{15}	3	5	5	2	5	4	5	5	1	W_1
X_{16}	3	5	5	3	5	4	5	5	1	W_1
X_{17}	3	5	5	4	5	4	5	5	1	W_1
X_{18}	3	5	5	4	5	5	5	5	4	W_4
X_{19}	3	5	5	5	1	5	1	5	5	W_5
X_{20}	3	5	5	5	1	5	5	5	5	W_5
X_{21}	3	5	5	5	5	4	5	5	1	W_1
X_{22}	3	5	5	5	5	5	5	5	4	W_4
X_{23}	4	2	1	5	1	5	1	5	5	W_5
X_{24}	4	2	5	5	1	5	1	5	5	W_5
X_{25}	5	2	1	5	1	5	1	5	5	W_5

3.5.2.2 Analysis of Importance of Parameters

If parameter A can well describe the change in parameter B, it can be said that A has a good classification ability for B, or A is greatly significant to B. Parameter A can be ignored if it is not essential or even not important at all to parameter B.

Based on discretization of parameters via Kohonen networks, it is necessary to analyze the importance of parameters a–h in Table 3.2 to parameter i. Let the status of a parameter i be W_1, W_2, W_4, W_5 according to the value of parameter i.

Le the 25 corresponding samples in Table 3.2 after reduction belong to the domain U. $U = (X_1, X_2, ..., X_{25})$. These samples are classified into four statuses according to the value of parameter i in the table.

$$W_1 = \left(X_2, X_3, X_4, X_5, X_6, X_7, X_9, X_{10}, X_{11}, X_{12}, X_{13}, X_{14}, X_{15}, X_{16}, X_{17}, X_{21}\right)$$
$$W_2 = \left(X_1\right)$$
$$W_4 = \left(X_{18}, X_{22}\right)$$
$$W_5 = \left(X_8, X_{19}, X_{20}, X_{23}, X_{24}, X_{25}\right)$$

Take W_1 as an example. The following is the analysis of classification ability of each parameter.

For parameter a, $a = \left(Y_1, Y_2, Y_3, Y_4, Y_5\right)$.

$$Y_1 = \left(X_1, X_2, X_3, X_4, X_5, X_6\right)$$
$$Y_2 = \left(X_7\right)$$
$$Y_3 = \left(X_8, ..., X_{22}\right)$$
$$Y_4 = \left(X_{23}, X_{24}\right)$$
$$Y_1 \cap W_1 \neq \Phi, Y_2 \subset W_1, Y_3 \cap W_1 \neq \Phi, Y_4 \not\subset W_1, Y_5 \not\subset W_1$$

So it can be concluded that $R_-(a) = Y_2 = (X_7)$

$$R^-(a) = \left(Y_1, Y_3\right) = \left(X_1, X_2, X_3, X_4, X_5, X_6, X_8, ..., X_{22}\right)$$
$$\alpha_a\left(W_1\right) = \frac{|U| - \left|R^-(a) - R_-(a)\right|}{|U|} = 0.2$$

Similarly, we can get $\alpha_a(W_2), \alpha_a(W_4), \alpha_a(W_5)$.

The importance of parameter $a, b, ..., h$ to the value of parameter i is shown in Table 3.3.

The mean values in Table 3.3 show that parameter f is the most important to i and parameter d is also important. Both of them should be retained. Some other parameters, such as parameter e and parameter g, are less important to i.

3.6 Data Correlation Analysis

During the testing of carrier rockets, it is very necessary to measure a great amount of variables for comprehensive performance evaluation and troubleshooting. Generally, there is some correlation between different variables. The change of such a correlation is an obstacle for related

Table 3.3 Importance of parameters.

	a	*b*	*c*	*d*	*e*	*f*	*g*	*h*
W_1	0.2	0	0	0.56	0	0.8	0	0
W_2	0.76	0.92	0.88	0.68	0.56	0.76	0.6	0.68
W_4	0.4	0.28	0.12	0.6	0.48	0.64	0.44	0.32
W_5	0.28	0.08	0	0.68	0.04	0.64	0.04	0.32
Mean value	0.41	0.32	0.25	0.63	0.27	0.71	0.27	0.33

staff to make an accurate judgment on the real reason that causes the change. Therefore, data correlation analysis is largely significant for performance evaluation and troubleshooting during the testing.

Based on testing data, data correlation analysis is for studying the correlation and the connection structure between different variables or variable-groups (including signals and data). Data correlation analysis can reveal the specific form and rule of variable change and decide how strong is the correlation between them. Correlation determination includes qualitative and quantitative analyses, which refer to make a judgment on whether there exists the correlation objective phenomenon and what relation it is based on theoretical knowledge and practical experience. Quantitative analysis is based on qualitative analysis, referring to make a judgment on direction, form and degree of the correlation by calculating the correlation.

Featured by small samples, nonlinearity, and the large scale, data correlation analysis for carrier rockets is different from that for some other engineering systems due to particularity and structural complexity of carrier rockets. Testing process is from bottom to top, for example, from part testing to subsystem testing to system testing, so analysis based on a priori knowledge and qualitative analysis of rocket system structure are a way to reveal the relation between variables and study their trend. Methods of qualitative analysis and description of correlation between variables include Petri's networks, signed directed graphs, and so on which will not be elaborated in this section. The focus of this section lies in quantitative analysis of correlation between variables, to be more specific, nonlinear correlation analysis based on kernel functions.

3.6.1 Algorithms and Measurements for Correlation Coefficients

3.6.1.1 The Concept of Correlation Coefficients

Correlation coefficient analysis is a classical method for quantitative analysis of data correlation. Correlation coefficient is an important statistical index represented by *r* for showing the direction and similarity of correlation between variables.

3.6.1.2 Algorithms for Correlation Coefficients

Let $X = (x_1, x_2, \ldots, x_n)^T$ and $Y = (y_1, y_2, \ldots, y_n)^T$ be the two sample waveform sequences for quantitatively describing the correlation, namely similarity, between X_t and Y_t, the two continuous variables. Translate them to a proper point. If one is a multiple of the other, they are 100% correlated; if one is sharply different from any multiple of the other or the two follow no rules, they are uncorrelated. Accordingly, the minimum error sum of squares can be adopted to measure their correlation.

$$Q_0 = \min_{a,\lambda} \frac{1}{n} \sum_{i=1}^{n} (y_i - a - \lambda x_i)^2 \tag{3.37}$$

If there exist certain a and λ that can get $Q_0 = 0$, it can be said that X_t and Y_t are 100% correlated. Otherwise, the value of Q_0 is used to describe their correlation. The value of Q_0 can be calculated by taking the derivative of the formula (3.38) with respect to a and λ

$$Q = \min_{a,\lambda} \frac{1}{n} \sum_{i=1}^{n} (y_i - a - \lambda x_i)^2 \tag{3.38}$$

Let the derivative be 0.

$$\frac{\partial Q}{\partial a} = \frac{-2}{n} \sum_{i=1}^{n} (y_i - a - \lambda x_i) = 0 \tag{3.39}$$

$$\frac{\partial Q}{\partial \lambda} = \frac{-2}{n} \sum_{i=1}^{n} \left[(y_i - a) x_i - \lambda x_i^2 \right] = 0 \tag{3.40}$$

We can get:

$$\lambda = \frac{\sum_{i=1}^{n} (x_i - \bar{X})(y_i - \bar{Y})}{\sum_{i=1}^{n} (x_i - \bar{X})^2}, \quad a = \bar{Y} - \lambda \bar{X}$$

Then we can get:

$$Q_0 = \frac{1}{n} \sum_{i=1}^{n} (y_i - \bar{Y})^2 \left\{ 1 - \frac{\left[\sum_{i=1}^{n} (x_i - \bar{X})(y_i - \bar{Y}) \right]^2}{\sum_{i=1}^{n} (x_i - \bar{X})^2 \sum_{i=1}^{n} (y_i - \bar{Y})^2} \right\} = \frac{1}{n} \sum_{i=1}^{n} (y_i - \bar{Y})^2 (1 - r_{xy}^2) \tag{3.41}$$

where

$$r_{xy} = \frac{\sum_{i=1}^{n} (x_i - \bar{X})(y_i - \bar{Y})}{\sum_{i=1}^{n} (x_i - \bar{X}) \sum_{i=1}^{n} (y_i - \bar{Y})} \tag{3.42}$$

We can also get the minimum relative error sum of squares:

$$E_0 = \frac{Q_0}{(1/n)\sum_{i=1}^{n}(y_i - \bar{Y})^2} = 1 - r_{xy}^2 \qquad (3.43)$$

E_0 can eliminate the effect brought by the measurement unit of X and Y, so it is more reasonable to use E_0 rather than Q_0 to measure the correlation between X and Y. To be more simplified, E_0 can be replaced by $|r_{xy}|$ to measure the correlation between X and Y, and $|r_{xy}|$ is called the correlation coefficient between X and Y. The greater the $|r_{xy}|$ and the smaller the E_0, the more closely X and Y are correlated; the smaller the $|r_{xy}|$ and the greater the E_0, the less closely X and Y are correlated. Sometimes r_{xy} is written as corr(X, Y) for convenience. Through the Schwarz inequality (3.44), an important nature of the correlation coefficient can be known: $0 \le |r_{xy}| \le 1$. Especially when $|r_{xy}| = 1$, X and Y have linear correlation; when $|r_{xy}| = 0$, X and Y are mutually orthogonal, namely the least correlated.

$$\sum_{i=1}^{n}(x_i - \bar{X})(y_i - \bar{Y}) \le \sqrt{\sum_{i=1}^{n}(x_i - \bar{X})^2 \sum_{i=1}^{n}(y_i - \bar{Y})^2} \qquad (3.44)$$

3.6.1.3 Measurements for Correlation Coefficients

The fundamental algorithm is the product moment method, PMM for short, developed by Karl Pearson, an English statistician. Pearson's correlation coefficient between two variables is defined as the covariance of the two variables divided by the product of their standard deviations. The form of the definition involves a "product moment," that is, the mean (the first moment about the origin) of the product of the mean-adjusted random variables; hence the modifier product-moment in the name.

$$r = \frac{S_{xy}^2}{S_x S_y} = \frac{\sum(x - \bar{x})(y - \bar{y})/n}{\sqrt{\sum(x - \bar{x})^2/n}\sqrt{\sum(y - \bar{y})^2/n}}$$
$$= \frac{n\sum xy - \sum x \sum y}{\sqrt{n\sum x^2 - (\sum x)^2}\sqrt{n\sum y^2 - (\sum y)^2}} \qquad (3.45)$$

The correlation coefficient is a relative number and an abstract statistical index. Its absolute value reflects that the degree of correlation of the absolute value is close to 1. The more closely the absolute value approaches 1, the higher the degree of correlation is; the more closely the absolute value approaches 0, the lower the degree of correlation is. The correlation coefficient, calculated through the mathematical formula, is represented by r which ranges from -1 to 1. The greater the absolute value of r, the more closely the variables are correlated; the smaller the absolute value of r, the less closely the variables are correlated. If $r > 0$ is positive correlation, then the two related variables have the same direction and if $r < 0$ is negative correlation, then the two similar variables have opposite direction.

3.6.2   Correlation Analysis by Principal Component Analysis

3.6.2.1   Correlation Analysis Based on Principal Component Analysis

The correlation coefficient reflects the degree of correlation between two variables. It will get so complicated if the correlation coefficients for all variables are to be analyzed, because there are a huge number of variables in rocket testing data.

Principal component analysis (PCA) is a typical method of feature extraction. It is a statistical procedure that uses orthogonal transform to convert a set of observations of possibly correlated variables into a set of values of linearly uncorrelated variables called principal components. The objective of PCA is to find out m orthogonal vectors as a set in data space. These vectors can reflect data variance to the best extent. Data can be mapped to the m-dimensional subspace made up of these orthogonal vectors from the original n-dimensional data space, and thus dimensionality reduction can be accomplished ($m < n$). PCA can be used to discover some major features in large samples and multiple variables and find out their correlation. Most importantly, original variables are seldom replaced by new ones in PCA. On the one hand, overlapping information can be eliminated; on the other hand, dimensions can be reduced. In this way, useful information can be selected among plane data. PCA simplifies multivariate data in the most effective and efficient way and accomplishes dimensionality reduction for high-dimensional data spaces.

For a data matrix, X, each column represents a vector, and each row represents sample observations.

$$X = TP = t_1 p_1^T + t_2 p_2^T + \cdots + t_a p_a^T \ (a < N) \tag{3.46}$$

t_i is the score vector and p_i is the load vector. Score vectors are pairwise orthogonal; load vectors are also pairwise orthogonal, and the module is 1. Each score vector is actually the projection of X at the same direction with its corresponding load vector, namely the principal component.

$$t_i = X p_i \tag{3.47}$$

The PCA of the matrix is virtually the eigenvector analysis of the covariance matrix. In practice, PCA may result in false variation and cannot display the true situation of data, because measurement units of process variables differ from one to another. To avoid this negative effect, it is necessary to eliminate the dimensional effect by normalizing original data.

Suppose $X = \left[x_{ij}^* \right]_{M \times N}$

$$x_{ij}^* = \frac{x_{ij} - \overline{x}_j}{s_j}, \quad i = 1,\ldots,M; \ j = 1,\ldots,N. \tag{3.48}$$

In the formula (3.48), $\overline{x}_j$ represents a mean value and s_j represents standard deviation. $p_i(i = 1,\ldots,N)$ stands for the change direction of data in X, and corresponding principal components are just the projection of data at this change direction. Caused by measurement noises, the change of X is mainly reflected by the direction of the first few load vectors instead of the last few ones on which the projection of X is quite small. Thus, the formula (3.46) can be revised as the formula (3.49).

$$X = t_1 p_1^T + t_2 p_2^T + \cdots + t_k p_k^T + E \tag{3.49}$$

In the formula, E is an error matrix, representing the change of X at the direction of the last $N - k(k < N)$ load vectors. In practice, E is largely caused by measurement errors, so ignoring E can eliminate the effect of measurement noises without losing useful information in data. In this way, dimensionality reduction can be well accomplished to guarantee effective analysis and evaluation in the testing process.

The non-zero mean value matrix can be centralized as the zero-mean matrix. The formula (3.50) represents the covariance matrix of (the zero-mean matrix).

$$C = \frac{1}{M} \sum_{i=1}^{M} x_i x_i^T \tag{3.50}$$

Solving the Equation 3.51 is just to make an eigenvector analysis on C.

$$\lambda_i p_i = C_i p_i \tag{3.51}$$

If ranking non-zero eigenvalues of C in a descending order, these eigenvalues' corresponding eigenvectors represented by p_i are just load vectors of X. By the formula (3.51), principal components of X can be computed. The cumulative explained degree of data is regarded as the basis for selection of specific principal components of y of the first k and that represented by the formula (3.52).

$$y = \frac{\sum_{i=1}^{k} \lambda_i}{\sum_{i=1}^{N} \lambda_i} \tag{3.52}$$

The essence of correlation analysis based on PCA is to reduce the number of dimensions to two for the system which is described by p-dimensional variables so that each sample point can be described in just one plane. In this way, it is easier to observe the correlation between sample points as well as distribution characteristics and structures of sample groups. PCA can make possible for the visibility of high-dimensional data points and enable system analyzers to timely discover universal laws and special phenomenon of large-scale and complex data groups, highly improving efficiency of data analysis. In terms of theories and methods of the decision support system, many famous experts and scholars have pointed out that transforming abstract high-dimensional invisible spaces or some more complicated spaces into visualized planar graphs can greatly promote insights of decision makers and increase their knowledge. It cannot be denied that correlation analysis based on PCA is one of the best ways to raise efficiency of decision support systems.

PCA can be applied for data correlation analysis during rocket testing, including data normalization, establishment of correlation coefficient matrixes, principal component calculation, and extraction.

Data Normalization
Let the length of a testing data sample be N, and the variable for correlation analysis be M. The original data matrix is

$$D = \left[d(1) d(2) \cdots d(N) \right] \in \mathfrak{R}^{M \times N}$$

$$d(t) = \left[d_1(t) \cdots d_M(t) \right]^T \in \mathfrak{R}^M$$

The aim of data normalization is to avoid unit inconformity and difference of order of magnitude between practical data. After normalization, sample mean is 0 and sample variance is 1. On this occasion, the covariance matrix is just the correlation matrix.

The data after normalization is

$$x_i(t) = \frac{d_i(t) - e(d_i)}{\delta(d_i)}, \quad \forall i, t, X = \left[x(1) \ x(2) \cdots x(N) \right] \in \mathfrak{R}^{M \times N} \tag{3.53}$$

where $e(d_i) = (1/N) \sum_{t=1}^{N} d_i(t)$ is the sample mean and $\delta^2(d_i) = [1/(N-1)] \sum_{t=1}^{N} [d_i(t) - e(d_i)]^2$ is the sample variance.

Establishment of Correlation Coefficient Matrixes
Establishment of correlation coefficient matrixes can reflect the correlation between samples. When all intersections are seen as one sample, the correlation coefficient matrix of these samples is

$$R(X) = \frac{1}{N-1} XX^T = \frac{1}{N-1} \sum_{t=1}^{N} x(t) x^T(t) = \left(r_{ij} \right)_{M \times N} \tag{3.54}$$

$r_{ij} = [1/(N-1)] \sum_{k=1}^{N} x_{ki} x_{kj}$ is the correlation coefficient of the ith and the jth variable. This correlation coefficient matrix reflects the correlation degree between M variables.

Principal Component Calculation
The correlation coefficient matrix R is applied to determine the eigenvalues ($\lambda_1 \geq \lambda_2 \geq \cdots \geq \lambda_M \geq 0$) and the corresponding orthogonal eigenvectors ($q(1), q(2), \ldots, q(M)$).

$$q(t) = \left[q_1(t) \cdots q_M(t) \right]^T \in \mathfrak{R}^M, \quad t = 1, 2, \ldots, M, \tag{3.55}$$

M principal components can be obtained:

$$y_i(t) = q(i)^T x(t), \quad i = 1, 2, \ldots, M. \tag{3.56}$$

The greater the variance, the more contribution it has to the total value. M subvectors of λ_i's corresponding eigenvectors $q(i)$ are the coefficients of M normalized variables in the ith principal component $y_i(t)$. Their absolute value and plus/minus sign reflects the correlation degree between this principal component and its corresponding variables.

Principal Component Extraction

In the application of PCA, only parts of them are needed for analysis. Suppose the first m are extracted as:

$$\left.\begin{array}{l} y_1(t) = q(1)^T x(t) \\ y_2(t) = q(2)^T x(t) \\ \quad\vdots \\ y_m(t) = q(m)^T x(t) \end{array}\right\}, \quad 1 \le m < N \qquad (3.57)$$

The accumulative contribution rate is $\Sigma\alpha_i$. Usually, if the accumulative contribution rate of the first m principal components is greater than or equal to 85%, the first m principal components can reflect the most information about all of the M variables..

Experiments have proved that it is the best to extract the first two principal components for the correlation analysis, because the M variables can be described in a two-dimensional space, which makes clearer the correlation between variables. In the matrix composed of principal components' corresponding eigenvectors, the greater the variance of the square value, the better it is. But in practical data processing, data may not be decentralized enough. To cope with this situation, the method of factor orthogonal rotation needs to be used. Eigenvalues of non-principal components are very small, so the factor model is based on principal components, which means that eigenvectors of the first m principal components constitute the loading matrix in the factor model. Orthogonal transformation can be conducted for the loading matrix, the purpose of which is to make data in the loading matrix more decentralized.

3.6.2.2   Kernel Function PCA

As a linear mapping algorithm, PCA is the theory of transform linear algebra. However, rocket working parameters are nonlinear, which multivariate statistical method may not work effectively. In typical nonlinear PCA, it has no way to get satisfactory results by a neural network algorithm, unless there are enough data samples. Therefore, the neural network algorithm is unsuitable for processing and analysis of rocket testing data. Kernel function principal component analysis (KPCA) is to map the input vector X to a high-dimensional space F through a previously selected nonlinear mapping. Thus, the input vector will have better separability. Making linear PCA on the mapped data in the high-dimensional space can get nonlinear principal components of the data. Some selected nonlinear principal components are seen as Eigen subspaces. KPCA is a typical approach in kernel function methods, suitable for handling small sample issues during rocket testing.

The basic idea of KPCA is to map the original input vector x to a high-dimensional eigenspace $\Phi(x)$ and then make a linear PCA calculation on $\Phi(x)$. The linear PCA for the eigenspace is equivalent to the nonlinear PCA for the input space.

Based on linear PCA, nonlinear PCA firstly introduces a nonlinear mapping $\Phi : x \in R^n \longmapsto \Phi(x) \in F \subseteq R^N$, and then transforms a sample set in the input space $x_k \left(x_k \in R^n, k = 1,2,\ldots M, \sum_{k=1}^{M} x_k = 0 \right)$ to the eigenspace (the number of dimensions of the eigenspace F could be arbitrarily large). Suppose $\sum_{k=1}^{M} \Phi(x_k) = 0$. The covariance matrix of samples in the eigenspace is

$$\bar{C} = \frac{1}{M}\sum_{i=1}^{M}\Phi(x_i)\Phi(x_i)^T \tag{3.58}$$

By solving the Equation 3.59, we can get the eigenvalue $\lambda(\lambda \geq 0)$ and its corresponding eigenvector $V(V \in F \setminus \{0\})$.

$$\lambda V = \bar{C}V \tag{3.59}$$

All the solutions for the equation above are in the subspace expanded by $\Phi(x_1), ..., \Phi(x_M)$, so we can get the Equation 3.60.

$$\lambda\left[\Phi(x_k)\cdot V\right] = \left[\Phi(x_k)\bar{C}V\right], k = 1, 2, ..., M \tag{3.60}$$

The coefficient $\alpha_i(i = 1, 2, ..., M)$ makes the formula (3.61).

$$V = \sum_{i=1}^{M}\alpha_i\Phi(x_i) \tag{3.61}$$

Combining the two formulas above, we can get

$$\lambda\sum_{i=1}^{M}\alpha_i\left[\Phi(x_k)\cdot\Phi(x_i)\right] = \frac{1}{M}\sum_{i=1}^{M}\alpha_i\left[\Phi(x_k)\cdot\sum_{j=1}^{M}\Phi(x_j)\right]\left[\Phi(x_j)\cdot\Phi(x_i)\right], \quad k = 1, 2, ..., M \tag{3.62}$$

Suppose there is a $M \times M$-dimensional matrix $K[K_{ij} = \Phi(x_i)\cdot\Phi(x_j)]$. We can get

$$M\lambda K\alpha = K^2\alpha$$
$$\alpha = (\alpha_1, ..., \alpha_M)^T \tag{3.63}$$

We can get nonzero eigenvalues by the characteristic Equation 3.64.

$$M\lambda\alpha = K\alpha \tag{3.64}$$

Due to that

$$1 = \sum_{i,j=1}^{M}\alpha_i^k\alpha_j^k\left[\Phi(x_i)\cdot\Phi(x_j)\right] = \sum_{i,j=1}^{M}\alpha_i^k\alpha_j^k K_{ij} = (\alpha^k\cdot K\alpha^k) = \lambda_k(\alpha^k\cdot\alpha^k) \tag{3.65}$$

α^k should be normalized to standardize the eigenvectors of the eigenspace, namely $(V^k\cdot V^k) = 1$. In order to extract principal components, it is necessary to calculate projections of samples in F on the eigenvector V^k. If x is the vector of a given test sample and its mapping in the eigenspace is $\Phi(x)$, the Equation 3.66 presents nonlinear principal components of the test sample corresponding to Φ.

$$\left[V^k \cdot \Phi(x) \right] = \sum_{i=1}^{M} \alpha_i^k \left[\Phi(x) \cdot \Phi(x) \right] \tag{3.66}$$

The discussion above is based on a precondition that all observed values are centralizing treatment. This precondition can be easily met in input spaces, but the mean value of samples after mapped into eigenspaces cannot be calculated exactly. One of the solutions to the problem is to replace K with $\bar{K}$ for getting eigenvalues. The formula (3.67) shows how to calculate $\bar{K}$.

$$\bar{K}_{ij} = K_{ij} - \frac{1}{M} \sum_{m=1}^{M} I_{im} K_{mj} - \frac{1}{M} \sum_{n=1}^{M} I_{in} K_{nj} + \frac{1}{M^2} I_{im} K_{mn} I_{nj} \tag{3.67}$$

It is worth mentioning that the largest number of KPCA principal components equals to M, the number of samples. So it can be said that KPCA principal components have a relation with samples. However, the number of linear PCA principal components depends on dimensionality of samples, which means linear PCA principal components have a relation with sample dimensionality. Similar to PCA, KPCA can also reduce the dimensionality of x if the first several eigenvectors are considered only. Through this kernel-based method, PCA nonlinearity can be realized, and $\Phi(x)$ of KPCA has the same characteristics as those of linear PCA. For KPCA, after utilizing the nonlinear mapping function $\Phi(x) \in R^M (M > N)$, the issue of nonlinearity in the input space will change into the issue in the high-dimensional eigenspace which can be solved by a linear method, namely kernel function, as shown in Figure 3.16. According to some functional theories, as long as a function $K(x,y)$ satisfies the Mercer condition, it has a corresponding interior product of a transformation space, namely $K(x, y) = \Phi(x) \cdot \Phi(y)$. Then, it is only necessary to carry out inner product calculation in the high-dimensional space. Such calculation can be conducted through the function in the original space, so there is no need to know the specific form of the space transformation. The most common kernel functions are linear kernel functions, polynomial kernel functions, Gaussian kernel functions, MLP kernel functions, etc. The Gaussian kernel function has been widely applied because of its simple algorithm, and it is obviously advantageous for solving complex problems.

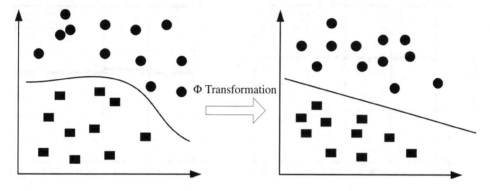

Figure 3.16 From linearly inseparable to linearly sortable.

3.6.3 Examples

Make an analysis on 10 sets of testing data. Testing signals are respectively A1, A2, A3, A4, A5, A6, A7, A8, A9, and A10, as shown in Figure 3.17

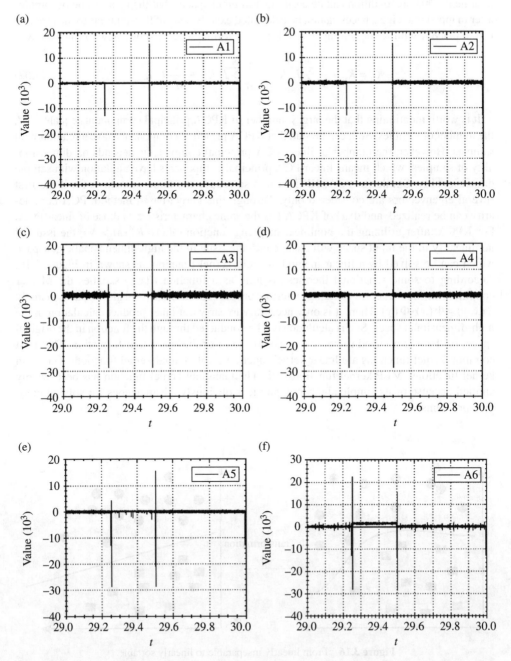

Figure 3.17 Testing signals. The data curve of the signal A1 (a), A2 (b), A3 (c), A4 (d), A5 (e), A6 (f), A7 (g), A8 (h), A9 (i), and A10 (j).

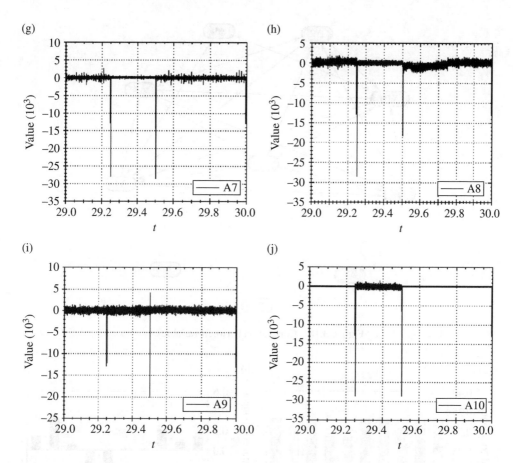

Figure 3.17 (*Continued*)

Table 3.4 Correlation coefficient matrix.

	A1	A2	A3	A4	A5	A6	A7	A8	A9	A10
A1	1	0.964	0.885	0.941	0.972	0.775	0.804	0.963	0.906	0.976
A2	0.964	1	0.865	0.923	0.954	0.797	0.789	0.942	0.866	0.954
A3	0.885	0.865	1	0.876	0.875	0.718	0.731	0.871	0.822	0.881
A4	0.941	0.923	0.876	1	0.945	0.800	0806	0.928	0.871	0.936
A5	0.972	0.954	0.875	0.945	1	0.829	0.801	0.955	0.899	0.966
A6	0.775	0.797	0.718	0.800	0.829	1	0.694	0.802	0.721	0.789
A7	0.804	0.789	0.731	0806	0.801	0.694	1	0.811	0.723	0.800
A8	0.963	0.942	0.871	0.928	0.955	0.802	0.811	1	0.894	0.956
A9	0.906	0.866	0.822	0.871	0.899	0.721	0.723	0.894	1	0.907
A10	0.976	0.954	0.881	0.936	0.966	0.789	0.800	0.956	0.907	1

According to the approach of PCA, the correlation coefficient matrix of A1, A2, A3, A4, A5, A6, A7, A8, A9, and A10 is shown in Table 3.4.

Based on the calculation results of the 10 signals' correlation coefficient matrix, let the associated threshold value be 0.9. The correlation of the 10 signals is shown in Figure 3.18.

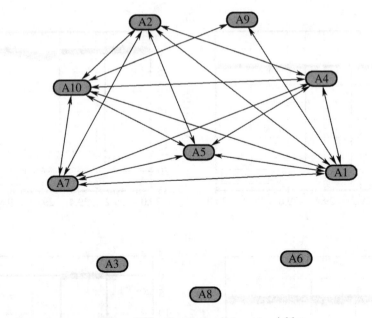

Figure 3.18   Correlation between variables.

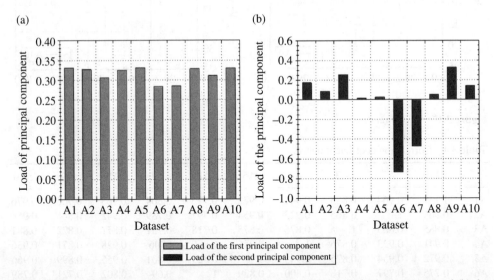

Figure 3.19   Load of principal components. The load of the (a) first principal component and (b) the second principal component.

It can be seen from Figure 3.18 that when the associated threshold value is 0.9, A3, A6, and A8 have no correlation with any other signal. The correlation coefficients between A2 and A9, A9 and A4, A9 and A5 are respectively 0.866, 0.871, and 0.899, all lower than 0.9. Accordingly, it is concluded that A2, A9, and A4 have no correlation with A5.

After transformation, there are two principal components whose loads are shown in Figures 3.19 and 3.20.

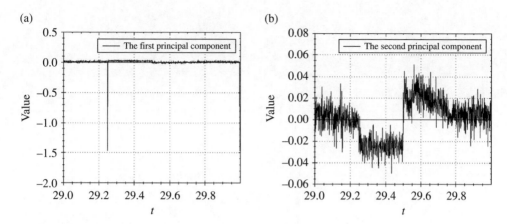

Figure 3.20 Principal component distribution. Output of the (a) first principal component and (b) second principal component.

Principal component load refers to the proportion of the original variable in the principal component. From Figure 3.19, it is known that in the first principal component, all variables have a large proportion because their correlation coefficients are all relatively high. In the second principal component, it is obvious that A6 and A8 have a large proportion, but in Table 3.4, A6 and A8 have a low correlation coefficient with other variables, which means A6 and A8 make up the second principal component independently. Meanwhile, it can be known from the output curves of the first and the second principal components that there is big difference between the variable outputs of the two principal components, so it is concluded that parts of A6 and A8 in the second principal component are highly independent.

4

Intelligent Fault Diagnosis for Space Launch and Testing

Fault diagnosis for space testing and launching includes two major parts: fault diagnosis for testing and fault diagnosis for launching. Fault diagnosis for testing refers to testing, diagnosis, location, and troubleshooting for the launch vehicle based on testing data during unit testing, subsystem testing, and sum-check testing. The objective of fault diagnosis for launching is to support security control analysis and decision-making, analyze the trend of parameters and working conditions based on telemetry data, diagnose abnormal situations, and conduct security assessment.

This chapter will analyze the problem of fault diagnosis for space testing and launching and discuss the development of intelligent fault diagnosis. The focus of this chapter is to discuss the combination of traditional fault diagnosis methods with intelligent ones such as the kernel principal component analysis (KPCA), wavelet analysis, ant colony algorithm, and neural network.

4.1 Overview of Fault Diagnosis for Space Launch and Testing

4.1.1 Cause Analysis

In space launches, the most fundamental cause of faults is the high complexity of launch vehicles, operating environments, and man-machine integration.

4.1.1.1 System Complexity

System complexity is the main cause of faults in space launch. Faults in large-scale space launches are inevitable because of the restriction of manufacturing technology. Conventional national and international space launches have proven that most faults of space launches can be attributed to defects or function failures of launch vehicle.

Intelligent Testing, Control and Decision-Making for Space Launch, First Edition. Yi Chai and Shangfu Li.
© 2015 National Defense Industry Press. Published 2015 by John Wiley & Sons Singapore Pte Ltd.

4.1.1.2 Process Complexity

Space testing and launching is an extremely complicated project. Test, control, and launch of the space vehicle like satellites, spacecrafts, space shuttles, and space stations show that most space vehicles break away the gravitational attraction into the space orbit by the constant and huge thrust provided by multistage rockets. No matter in the takeoff phase, the powered flight phase, the attitude control phase, the orbit supervising phase, or the recovery phase, the carrier rockets have to rely on the monitoring, control and intervention of testing and control network in the ground or low-altitude; however, the control instruction of testing and control network works through the cooperation of equipments on the rocket (satellite). On the one hand, monitoring, control, and intervention are all remote operation methods, such as telemetry and telecontrol. Therefore, the fault or mistakes of any part (such as measuring, calculating, deducing, decision-making, controlling, and signal emitting) will lead to the systemic failure or disaster. On the other hand, working in coordination and cooperation, each link or node in the network system must cooperate with each other in the whole process with clear instructions, stable signals, and unified time. Any fault or function failure of any network node will lead to failure of the system, even the whole project.

4.1.1.3 Technology Complexity

Space launch is an area with high risks due to technology complexity, which is reflected in two aspects. Firstly, space launch has the characteristics of the integration of multi-technology and the comprehension of multi-disciplinary; secondly, target diversification as well as complexity of structures and functions of the space vehicles is required increasingly in the military and civil fields. Practice has proven that some newly invented materials and technologies, which are stable on ground or in simulated space environment may be unreliable and ineffective or even have potential problems when they are utilized in space projects.

4.1.1.4 Environment Complexity

During space launches, space carriers and vehicle are running in an extremely complex environment. Space carriers have to undertake various interferences when they are carrying space vehicle to traverse the dense atmosphere, the ionosphere, and the magnetosphere. For both mechanical and electronic equipments, their functions, performance, and stability will experience massive changes compared with ground design and testing, which will trigger a variety of abnormal phenomenon inevitably. It can be said that either environmental uncertainty or climatic anomalous changes may become the fault source in space projects. Practice has proven that complex environments can not only lower the system reliability but also affect the safe operation of the system directly.

4.1.1.5 Man–Machine System Complexity

From the perspective of man–machine engineering, space launch is a structurally complicated system involving many people, machines, and environments. In this system, working staff directly or indirectly control rocket launch, testing, tracking, orbital maneuver, attitude and rocket security, so the human factor is also an important contributing factor that causes faults. In the testing, launching, control, and management of space launch, many emergency situations and subjective or objective factors, such as operators' psychology, physiology, knowledge, and experience, will make the operators send false instructions and act in a wrong way, or make wrong judgments and decisions.

4.1.2 Characteristics

Space testing is a super complicated system. It has not only the typical characteristics that normally complicated systems have, such as autonomy, uncertainty, development, and dispersibility, but also the characteristics that other complicated systems do not have, such as nonrepeatability, irreversibility, singularity, strong correlation between systems and environments, noncontinuity, and singularity of the changing of process parameters. Space faults not only have the characteristics of hierarchy, transmissibility, correlation, radioactivity, and time delay that normally complicated systems have but also the following characteristics due to the particularity of space systems.

4.1.2.1 Multi-Occurrence of Faults

High incidence rate of space faults refers to a high percentage of tests where faults occur in the total number of tests during a space project, and it also means multiple faults in a single test. Reports said that from 1957 to 1994, there were 136 failures out of 1043 launches of liquid-propellant rockets in the Unites States that involved Vanguard, Juno, Thor, Delta, Atlas, and Titan, where the failure rate reached 13.04%; there were 47 failures out of 1220 rocket launches in the Soviet Union that involved Moon, Vostok, Soyuz, Proton, Cosmos, and Zenit, where the failure rate reached 3.85%; as to European Space Agency, there were 11 failures out of 126 launches of the Ariane family, where the failure rate reached 8.73%; there were 27 failures in Japan that involved the families of K, L, M, and H.

4.1.2.2 Concurrency of Multiple Faults

Space testing, launch, and control is a complicated system that involves multiple people, machines, environments, and structures. It can be divided into several subsystems—testing and launch system, tracking, telemetry and command system, communication system, meteorology system, and service support system, each of these subsystems consisting of their own subsystems. It is acknowledgeable that interrupt and fault will happen in any subsystem and node in the process of testing, launch, control, and management. Therefore, fault concurrency is comparatively common in space projects. Fault concurrency means that multiple failures occur simultaneously or successively during a short period of time or that a single fault results in failures of various subsystems at one time point.

4.1.2.3 Diversity of Fault Modes

Due to the strong coupling between the subsystems, diversity of fault modes refers to that different subsystems, different segments, and different types of faults are associated with each other. Generally, in a subsystem, sudden failures with complex impacts may occur simultaneously with continuity, infectivity, and leakage faults. A small fault may result in a big disaster. One single fault tends to be the cause of another fault in different subsystems.

4.1.2.4   Irreparability and Irreversibility

Most faults, perhaps control system faults, engine faults, or stage separation faults, are irreparable during the process of launch and flight, although there are timely detection and diagnosis. The only possible mean to be taken is flight monitoring, judgments of fault consequences, and determining the security control measures.

4.1.2.5   Harmfulness of Fault

The cost of human, financial, and material resources for each large-scale space experiment is usually more than 100 million or even over 10 billion dollars. If the launch fails, there will be a huge economic loss. Furthermore, the price will be higher if the failed launch poses a threat to the launch area, the flight area, and even the environment that people rely on.

4.1.3   Difficulties

The complexity of space systems makes space fault analysis and diagnosis a difficult and challenging research area in space testing, launch and control.

4.1.3.1   High Reliability and Accuracy

Space launch is required to be 100% secure and reliable because of its characteristics of irreversibility, part non-reusability, and strong correlation between systems and environments. Not only each launch is required to be secure, but fault diagnosis is also required to be able to describe the change trend and the law of development of reliability. Related maintenance measures should be taken to predict and discover faults and to prevent disastrous consequences caused by sudden faults.

4.1.3.2   Technological Comprehensiveness

All systems in space launch projects are systemized, large scaled, complicated and technologically advanced. With rapid development of modern information technology, more and more advanced technologies have been utilized in space vehicle, carrier rockets, ballistic missiles and launch sites. These newly developed technologies and materials include microelectronic technology, photoelectric technology, computer networking technology, artificial intelligence technology, advanced composite materials, stealth materials, and heat-resisting materials which makes system hardware more complicated and software more concentrated. Therefore, space fault diagnosis is increasingly technologically comprehensive.

4.1.3.3   Quick Response

Fault process in a space launch is supposed to be timely and prompt. The difficulty of fault diagnosis in space testing and launch lies on quick fault discovery, location, and response.

4.1.3.4   Lack of Prior Information

Space launch is an exploratory and experimental activity. The number of launches of space vehicle, carrier rockets, and ballistic missiles is no more than 20 each year, and each launch is with new experimental objectives and technical conditions. Moreover, the number of testing for products in development is also limited. Therefore, the pretest information for space fault diagnosis is very insufficient, which makes fault diagnosis more difficult.

4.1.3.5   Risk of Decision-Making in Fault Diagnosis

It is quite risky to test, judge, and handle faults in a wrong way. For example, in the carrier rockets take-off stage, misjudging a normal flight as a faulted one may result in an explosion of the rocket and its payload, which will cause a huge economic loss. Conversely, if the faulted rocket that should have been exploded in a secure way falls down, the environment will be largely polluted and worse still, it may cause casualties.

4.2   Fault Diagnosis Methods

4.2.1   Classification of Fault Diagnosis

There are two categories of fault diagnosis methods—qualitative analysis and quantitative analysis.

4.2.1.1   Qualitative Analysis

The method of qualitative analysis consists of graph theories, expert systems, and qualitative simulation. Graph theories include signed directed graph (SDG) and fault tree.

1. SDG is a kind of graphical mode, which is widely adopted to describe system causality. In SDG, incidences or variables are represented by nodes, and the causality between variables is represented by lines with the direction from cause nodes to result nodes. When the system is normal, nodes in SDG are also in normal conditions; if some faults emerge, the faulted nodes will deviate from their nominal values and give an alarm. According to the causality between nodes in SDG as well as some search strategies, all possible fault propagation paths and their evolution process can be analyzed. Besides, the reason why faults happen can be clarified too. Fault tree is a special logic figure. The diagnosis method based on a fault tree is an analyzing process from the result to the cause. Such method begins with fault condition and then goes to inferential analysis. Finally, basic causes, impacts, and incidence of faults will be found out. Fault diagnosis methods underlying graph theories due to the characteristics of easy modeling and accessible results are widely used in practical application. However, these methods will be extremely difficult for fault diagnosis when the system is comparatively complicated. The accuracy may not be satisfactory because invalid diagnosis results may be presented.
2. Fault diagnosis methods based on expert systems refer to building knowledge bases and designing a set of computer programs. The knowledge bases are the experience of domain experts, which accumulated in the long-term practice. The computer program is utilized to

simulate human inference, decision–making, and diagnosis. Expert systems mainly consist of knowledge bases, inference engines, integrated data bases, man–machine interfaces, and interpretive modules. Generally, expert knowledge is inevitably uncertain. To solve this problem, fuzzy expert systems is proposed where the concept of fuzzy membership degree is explored and the fuzzy logic is applied to do the reasoning. Fault diagnosis methods based on expert systems have been widely applied because rich expert experience and knowledge are used and results are easy to understand without mathematical modeling. However, such methods have their own demerits. Firstly, knowledge is difficult to be obtain, which becomes the major bottleneck in expert system development. Secondly, diagnosis accuracy depends on the richness of expert experience and the proficiency of expert knowledge in knowledge bases. Finally, when there are too many rules, there will exist many problems such as matching conflicts and combinatorial explosions, which will slow down inference and lower efficiency.

3. Qualitative simulation is to describe the qualitative behavior of the system. In normal or faulted circumstances, qualitative behavior description of the system in qualitative simulation can be seen as system knowledge for fault diagnosis. The qualitative simulation method depending on qualitative differential equations is one of the most mature research methods. In such method, the system is described as a symbol set of physical parameters. The method is a constraint equation set reflecting the correlation between these physical parameters. Starting from the original state, systems have various subsequent states, and constraint equations are used to filter those unreasonable states. This process will repeat until there are no new states emerging. The most obvious characteristic of qualitative simulation is that it can help do the reasoning on dynamic behavior of systems.

4.2.1.2 Quantitative Analysis

Quantitative analysis mainly consists of methods based on analytical models or data driving.

1. Analytical models contain two parts, based on accurate mathematical models and observable inputs and outputs structuring. First residual signals are constructed to reflect the disparity between expected system behavior and practical operating mode. And then fault is detected according to the analysis on residual signals. The research of fault diagnosis based on analytical models is comparatively mature and profound. In general, fault diagnosis methods based on analytical models include state estimation, parameter estimation, and parity space.

   Fault diagnosis methods based on state estimation mainly include the filter method and the viewer method. Fault diagnosis based on parameter estimation suggests that fault diagnosis can be conducted through the detection of parameters in models. Because faults will cause changes in system procedure parameters and such changes will further lead to changes in model parameters. Fault diagnosis methods based on parity space establish equivalent mathematical relationships between input variables and output variables by using systematic analytic mathematic models. It reflects static direct redundancy among output variables and dynamical analytical redundancy between input variables and output variables. Verifying whether practical input and output values satisfy the equivalence relation can fulfill the goal of detecting and separating faults. Fault diagnosis methods based on analytical models are very effective due to the profound understanding on internal parts of systems. These categories of methods rely on accurate mathematical models of diagnostic subjects, but practically it is very difficult to build accurate mathematical models of diagnostic subjects. Accordingly, these fault diagnosis methods based on analytical models are no longer suitable. However, a huge number of data are accumulated during the system operation, so fault diagnosis methods based on process data are imperative for fault diagnosis.

2. Fault diagnosis methods based on data driving analyze and deal with process data. Such fault diagnosis can be accomplished without accurate analytical models. These fault diagnosis methods based on data driving includes machine learning, multivariate statistical analysis, signal processing, information fusion, and rough sets.

   Fault diagnosis methods based on machine learning use systems' historical data in normal and other conditions to train neural networks or support vector machines (SVM) for fault diagnosis. These methods are widely applied and focus on the accuracy of fault diagnosis. However, machine learning algorithms need sample data in faulted circumstances, and the accuracy is greatly related to completeness and representativeness of samples. Therefore, these methods are not effective because it is impossible to collect a huge number of fault data.

   Fault methods based on multivariate statistical analysis work by applying correlation between variables. In this category of methods, the approach of multivariate projection is based on historical data. The multivariable sample space is decomposed as a low-dimensional projection subspace expanded by principal variables and a corresponding residual subspace.

   In these two subspaces, statistics that can reflect space changes are to be worked out, and then observation vectors are to be projected into the two subspaces respectively. The corresponding statistic index will be used in process monitoring. Common monitoring statistics include T^2 in projection spaces, Q in residual spaces, Hawkins, and global Markov distance. For fault methods based on multivariate statistical analysis, it is not necessary to get a deep understanding on system structures and principles. What really matters is the testing data obtained during system operation, because the algorithm is simple. However, it is difficult to explain the faults diagnosed by such methods, and some issues in these methods are still needed to be solved, such as nonlinearity of process variables, process dynamics, and time-dependent nature, due to the complexity of practical systems.

   Fault diagnosis methods based on signal processing refer to analyzing measurement signals through various signal processing approaches and extracting time and frequency domains features related to faults for diagnosis. Such methods include spectrum analysis and wavelet transformation.

   Rough set is a new mathematical tool for discovering knowledge from data and revealing its potential rules. It is quite different from fuzzy theory based on subordinating degree functions and evidence theory based on confidence coefficient. A great advantage of rough set is that only data sets, rather than any other prior subjective information, are needed to describe and process uncertain situations objectively. Attribute reduction is the core of rough set theory. Correct classification results are obtained from the least attribute information by deleting unrelated or unimportant condition attributes without affecting system decision-making. Thus, rough set can be used to select the effective fault feature to set in fault diagnosis in order to reduce the number of input eigenvector dimensions and downsize fault diagnosis systems.

Information fusion technology can help get more reliable conclusions by automatically analyzing and integrating multisource information than single-source information. Fault diagnosis methods based on information fusion rely partly on complementation and redundancy information of multisensors. How to secure the effective utilization of this information is still an opening problem, which is aimed at improving the accuracy of fault diagnosis and reducing the number of false reports.

In conclusion, fault diagnosis methods based on data driving do not need accurate analytical models. Such methods are more easily to be applied in practical systems because they depend only on historical system data. However, these methods are not competent for fault analysis and explanation, because systems' internal structure and mechanism information are not available.

4.2.2   Intelligent Fault Diagnosis Methods

In recent years, artificial intelligence and computer technologies have greatly developed. Many new theories and technologies make various intelligent fault diagnosis methods possible. Intelligent fault diagnosis includes two aspects. The first is to apply artificial intelligence directly into fault diagnosis, including expert system and machine learning. As mentioned above, expert system has the advantage of knowledge representation and machine learning achieves fault identification and diagnosis by using neural networks and SVMs. The second is to combine artificial intelligence and unintelligent methods. Artificial intelligence technologies can overcome shortcomings and improve the inadequacy of traditional fault diagnosis methods. There are three ways of combination.

1. Artificial intelligence is combined with expert systems, fuzzy theories, rough sets, and wavelet analysis to solve the issue of lacking accurate mathematical models.
2. Artificial intelligence is combined with neural networks and SVMs (including kernel functions) to solve diagnosis issues involving nonlinearity and complicated classification. The methods linked with SVMs (including kernel functions) can also be applied to small sample subjects, such as multivariate statistical analysis based on SVMs and sneak circuit analysis (SCA) based on neural networks.
3. Artificial intelligence is combined with genetic algorithm, immune algorithms, and swarm intelligence algorithm to address the issues of complex optimization in diagnosis. Graph methods based on the ant colony algorithm are just one example.

4.2.3   Intelligent Fault Diagnosis for Space Testing and Launching

In current space testing and launching, some faults are not caused by complicated factors, so common fault diagnosis technologies turn out to be effective. In last 50 years, China has developed a number of common fault diagnosis methods from practical experience and accumulation. It can be subdivided into fault tracking and tracing, fault isolation checkout, status checkout, physical examination, environmental testing, correlation method, and fault simulation examination. However, for some complicated and comprehensive faults, these methods are not capable enough. More effective fault diagnosis methods are most urgently needed to be studied and developed.

After a long period of practice and accumulation, researchers should combine artificial intelligence with traditional fault diagnosis methods to develop diagnosis solutions for complicated faults in space testing and launching. Firstly, multiscale principal component analysis (MSPCA) is introduced for feature extraction, in which, the intelligent multiscale principal statistical analysis based on data driving is discussed. Secondly, faults can be located through intelligent fault tracking and searching depending on the correlation between system state and its faults. Finally, the potential path relation of faults is described through neural networks.

4.3   Multiscale Fault Detection Algorithms Based on Kernel PCA

As an information-processing method based on data driving, PCA is widely used in the data correlation analysis and fault diagnosis. The core of PCA is to make statistical analysis based on the correlation between multiple process variables. The essence of PCA is to reduce dimensions of a high-dimensional data space. And then a hidden variable space is established to replace the original variable space through multivariate projection. It is composed of a low-dimensional projection subspace expanded by principal component variables and a corresponding residual subspace. Afterward, the various functions will be realized in this space. To be more specific, the first step is to construct statistics that can reflect corresponding space changes in the principal component space and the residual space and then to project observation vectors into the principal component space and the residual space respectively. Inspired by the practical monitoring statistics, we can know the state of the system. If the practical monitoring states exceed the previously-set process monitoring indicators, the system is faulty. Common monitoring statistics include T^2 in projection spaces, Q in residual spaces, Hawkins, and global Markov distance.

Conventional PCA is based on fixed, single scaled and linear analysis, while in practice, diagnosis subjects are usually nonlinear and the data collected in practical situations are always accompanied with noises and interference. Diagnosis subjects are in different scales and changes over the time. Moreover, some complicated systems, such as the carrier vehicle system, also have the small sample characteristics due to the limited number of data samples. Therefore, PCA is not the best method for the practical application of process data-based models.

To make up for the deficiency of traditional PCA, researchers at home and abroad have presented the MSPCA and adaptive principal component analysis (APCA) such as KPCA, moving window principal component analysis (MWPCA), and recursive principal component analysis (RPCA).

As mentioned in Chapter 3, kernel PCA is an approach based on kernel functions, in which input vectors are mapped nonlinearly into a high-dimensional feature space. In this way, input vectors can enjoy better separability and nonlinear component extraction can be accomplished under small sample condition. APCA can be used to monitor equipment performance or newly updated situations according to mean value, variance, covariance, and model control of process data. APCA can also effectively deal with slowly varying and mutational variables. Besides, it can contribute to make avoidable data time-variant effect and model deviation caused by equipment degradation. However, it is not considered that data's multiscale feature is caused by events that happen at a different time, in different frequency, and at different locations in APCA. MSPCA can effectively deal with different kinds of nonstationary process data and explain the cumulative effect from different angles so that it can monitor process variables accurately. Besides, MSPCA obtains various statistic features of process data so that to prevent delayed or misleading fault reports.

4.3.1 Principles of PCA Based on Fault Diagnosis

PCA is an effective approach to transform multiple related variables into several mutually-independent variables. The ultimate goal of PCA is to find key data which can reflect original data information or to reduce dimensionality for a high-dimensional space. Namely, PCA can extract representative statistic features through compression of the original data space. As a typical statistical process control theory, PCA achieves performance supervision and fault diagnosis under normal circumstances. First, principal component models are established to directly extract the correlation between historical process data. Then system state is judged by examining the extent of the deviation of newly observed data and historical process data of statistic model.

After the establishment of PCA models based on historical process data under the normal systematic condition, controlled variables in multivariate statistics can be applied to diagnose, analyze, and detect faults. Commonly used statistical magnitudes are Hotelling T^2 and squared prediction error (SPE).

Definition 4.1   Hotelling T^2 in the principal component subspace

$$T_i^2 = \sum_{j=1}^{m} \frac{t_{ij}^2}{S_{t_i}^2} \tag{4.1}$$

In Equation 4.1, T_i^2 is the ith row of T^2, m is the number of selected principal components, t_{ij} is the jth row of t_i, and $S_{t_i}^2$ is the estimated variance of t_i. The control limit is determined by F distribute.

$$\text{UCL} = \frac{k(n-1)}{n-k} F_{k,n-1,\alpha} \tag{4.2}$$

In Equation 4.2, n is the sample number of principal component models, k is the number of selected principal components, α is the test level, and $F_{k,n-1,\alpha}$ is the critical value of F distribute when degree of freedom is k and $n-1$ respectively.

Definition 4.2   SPE in the residual subspace

$$\text{SPE}(k) = \left\| E(k)^2 \right\| = x(k)(I - P_t P_t^T) x(k)^T \tag{4.3}$$

In Equation 4.3, P_t is the data matrix formed by the first t lines of the corresponding loading matrix in the principal component model. The control limit of SPE is determined by corresponding normal distribution:

$$Q_\alpha = \theta_1 \left[\frac{h_0 C_\alpha \sqrt{2\theta_2}}{\theta_1} + \frac{\theta_2 h_0 (h_0 - 1)}{\theta_1} + 1 \right]^{\frac{1}{h_0}} \tag{4.4}$$

$$\theta_j = \sum_{i=t+1}^{T} \lambda_i^j, \quad j = 1, 2, 3, \tag{4.5}$$

$$h_0 = 1 - \frac{2\theta_1 \theta_3}{3\theta_2^2} \tag{4.6}$$

where C_α is α quantile of the normal distribution.

SPE and T^2 respectively reflect the change situation from different perspectives of partial observation data explained by the principal component models. SPE is the deviation degree of $x(k)$-observation data at the time of k compared with its principal component model. It is used to describe the change trend of the principal component model's corresponding external data. T^2 is the degree of deviation of each sampling data in the amplitude value and changing trend compared with selected principal component models. This degree of deviation is regarded as an important measure to evaluate the changing situation inside principal component models.

PCA can be applied to detect whether practical systems are in normal conditions through testing whether T^2 and SPE exceed their own corresponding control limit. If collected online data and the data of principal component models construction are all in normal conditions, corresponding T^2 and SPE in principal component models will both be lower than the T^2 and SPE limits set in PCA models. Otherwise, T^2 and SPE will exceed the previously-set control limits.

Take $x = [x_1, x_2, x_3, x_4]^T$, a four-dimensional multivariable system as an example. The system's observability matrix $X \in R^{4 \times 1990}$ can be represented as Equation 4.7.

$$X = \left[x(1), x(2), ..., x(1990) \right] \tag{4.7}$$

In $X \in R^{4 \times 1990}$, 4 represents the number of system state variables and 1990 represents the number of sampling times.

$$\begin{cases} x_1 = \sum_{i=3}^{7} \frac{1}{2} \sin \left\{ 2\pi \left[2^{i+1} + \text{rand}\left(2^i\right) \right] \bar{t} \right\}, & 0 \le \bar{t} \le 1 \\ x_2 = N(0,1) \\ x_3 = x_1 + x_2 \\ x_4 = x_1 - x_2 \end{cases} \tag{4.8}$$

It is impossible to avoid the noise of measured variables, so to promote experimental authenticity, white Gaussian noises are introduced into systems whose mean value is 0 and variance is 1, which is shown in Equation 4.9.

$$\bar{X} = X + 0.1N(0,1) \tag{4.9}$$

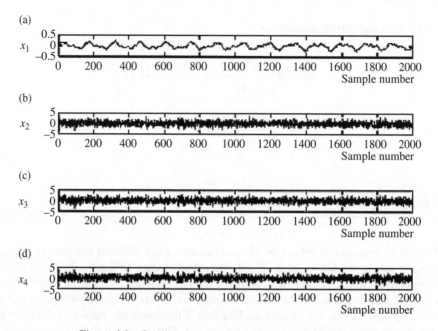

Figure 4.1   Original signals: (a) x_1, (b) x_2, (c) x_3, and (d) x_4.

After high-frequency faults are added in the section from 800 to 1000, we can get the signals shown as in Figures 4.1 and 4.2.

Analyze the system by PCA and select the number of principal components through the accumulative contribution rate. After calculation, two principal components are selected in this case. The first principal component has the data interpretation degree of 73.4546% and the second principal component 25.1135%. The data description degree of the two principal components is 98.5681%. Figure 4.3 shows the accumulation interpretation degree of the two principal components in the model in this case.

Each principal component in principal component models is virtually the projection of the data matrix $\bar{X}$ at the direction of the principal component's corresponding load vector. Such space is called a principal component space. Figure 4.4 shows two principal components' location in a plane (a hyperplane actually). The abscissa refers to the first principal component vector and the ordinate refers to the second principal component vector.

Although Figure 4.4 describes the location of the two principal components in the principal component space, it does not display the reason why the two principal components have been selected. The length of the principal component t_i reflects the coverage degree of the data matrix $\bar{X}$ at the direction of the load vector p_i. The longer the length is, the larger the coverage degree will be. If ranking principal components according to their length $\left(\|t_1\| > \|t_2\| > \cdots > \|t_m\| \right)$, the load vector p_1 refers to the direction with the biggest data changed in the data matrix $\bar{X}$, and p_2, vertical with p_1, refers to the direction with the second biggest data change in the data matrix $\bar{X}$. Similarly, p_m, vertical with p_{m-1}, refers to the direction with the least big data change in $\bar{X}$. Figure 4.5 intuitively describes the projection

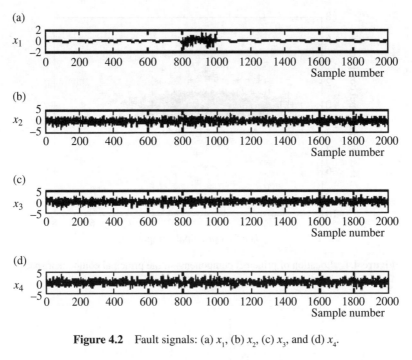

Figure 4.2   Fault signals: (a) x_1, (b) x_2, (c) x_3, and (d) x_4.

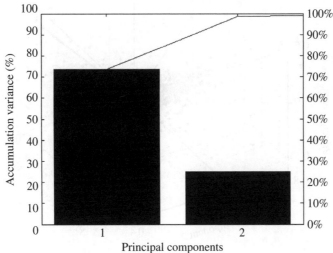

Figure 4.3   The accumulation interpretation degree of principal components on original data.

situations of four principal components $x_1 \sim x_4$ in the principal component space. It can be seen that x_1 and x_2 have the highest data coverage degree, while x_3 and x_4 have the lowest one. Therefore, x_1 and x_2 are selected as the two principal components in this case. This also happens to verify the discipline that if certain linear correlation exists between variables in $\bar{X}$, the main data change in $\bar{X}$ is reflected by the first several load vectors.

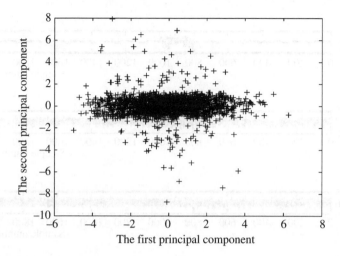

Figure 4.4   Location of principal components in the principal component space.

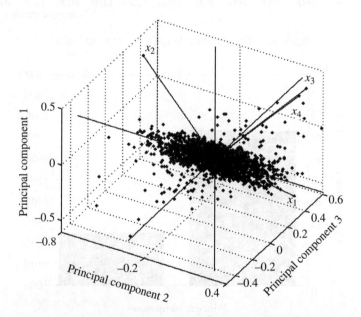

Figure 4.5   Projections of vectors in the principal component space.

In this case, control limit of T^2 and SPE is 95%. Through analysis, we can get T^2 and SPE monitoring figures as shown in Figures 4.6 and 4.7.

Figures 4.6 and 4.7 are displayed that both T^2 and SPE monitoring figures detect the faults from the region 800 to 1000. However, misinformation can also be seen in either of

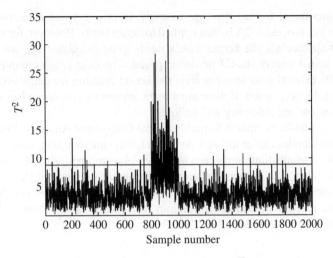

Figure 4.6 T^2 monitoring diagram of $\overline{X}$ (PCA).

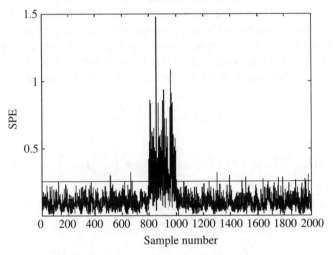

Figure 4.7 SPE monitoring diagram of $\overline{X}$ (PCA).

the figures. In SPE monitoring figure, there is less misinformation than that in T^2 monitoring figure. In conclusion, it is not the best way to utilize PCA with a single scale and a fixed model for fault diagnosis.

4.3.2 KPCA Fault Detection Based on Feature Sample Extraction

Standard KPCA fault detection approaches, which is similar to the idea of KPCA correlation which has been discussed in Chapter 3, has introduced kernel functions into PCA fault diagnosis. The basic idea of KPCA is to select a reasonable kernel function and map input vectors into a high-dimensional feature space nonlinearly. In this way, input vectors can be more divisible. After this process, PCA is then applied to detect faults. However, for standard KPCA fault detection approaches, the kernel matrix needs to be calculated and stored, and eigenvalues of the kernel matrix should be decomposed. The calculation complexity is $O(M^3)$. Moreover, while extracting sample features, the kernel function for calculation samples and all training samples is required. If there are a huge amount of samples to be extracted, it will cost a lot of time and the efficiency will be low.

To avoid these problems, Sparse Kernel Principal Component Analysis (SKPCA) which is based on feature samples can be utilized. At present, there are two categories of solutions to K calculation. One is to substitute some data in the kernel matrix with zero, thus to form a sparse matrix, and the other is to reduce the number of training samples. SKPCA is an approach depending on sample feature extraction. The extraction of samples does not simply mean the reduction of the number of samples at random but to keep sample distribution unchanged through feature sample extraction.

4.3.2.1 The Principle of Sample Feature Extraction

The image of the original data x_i in the mapping space F is represented by $\phi(x_i)$. Suppose $\phi_i = \phi(x_i)$ and $k_{ij} = \phi_i^T \phi_j$. The feature sample selected out of N samples is $X_s = \{x_{s1}, ..., x_{sL}\}$. The projection of other samples in the mapping space F can be represented by the approximate mapping of the feature sample, namely $\hat{\phi}_i = \varphi_s \cdot a_i$, where $\varphi_s = (\phi_{s1}, ..., \phi_{sL})$ and $a_i = (a_{i1}, ..., a_{iL})^T$. a_i is the coefficient vector that leads to the smallest difference between $\hat{\phi}_i$ and ϕ_i. The difference between $\hat{\phi}_i$ and ϕ_i can be represented by $\delta_i = \left\| \phi_i - \hat{\phi}_i^2 \right\| / \left\| \phi_i^2 \right\|$.

$$\min_{a_i} \delta_i = 1 - \frac{K_{si}^T K_{ss}^{-1} K_{si}}{K_{ii}} \tag{4.10}$$

In which, $K_{ss} = \left(k_{s_p s_q} \right)$, $1 \le s_p \le L$, $1 \le s_q \le L$, $k_{s_p s_q} = \phi^T \left(x_{s_p} \right) \phi \left(x_{s_q} \right)$. x_{s_p} and x_{s_q} are feature samples. $K_{si} = \left(k_{s_p i} \right)$, $1 \le p \le L$.

The extracted feature sample set S is supposed to satisfy the requests of representative index, so the sum of δ_i should be minimized.

$$\min_S \left[\sum_{x_i \in X} \left(1 - \frac{K_{si}^T K_{ss}^{-1} K_{si}}{K_{ii}} \right) \right], \quad \max_S \left[\sum_{x_i \in X} \frac{K_{si}^T K_{ss}^{-1} K_{si}}{K_{ii}} \right] \tag{4.11}$$

$J_s = (1/N) \sum_{x_i \in X} J_{si}$, and $J_{si} = \left(K_{si}^T K_{ss}^{-1} K_{si} \right) / K_{ii} = \left\| \hat{\phi}_i^2 \right\| / \left\| \phi_i^2 \right\|$. Therefore, Equation 4.11 is equal to $\max_S (J_s)$. It can be seen that the value range of J_s and J_{si} is $(0, 1]$.

4.3.2.2 The Algorithm of Sample Feature Extraction

The algorithm of feature sample extraction is a cyclic process. Initially, middle sample of the sample set is extracted. There is only one single sample ($L=1$) in the feature sample set S. Calculate the fitness of S, namely J_s and J_{si}. The corresponding sample of the smallest J_{si} is to be added into S. In the second step, calculate the fitness of the new feature sample set S.

The cyclic process will not stop until the value of J_s meets the requirement. The steps for the algorithm of feature sample extraction are as follows:

1. Determine a stopping criterion, namely the max Fitness.
2. Extract middle samples x_m, $S = \{x_m\}$, $L = 1$.
3. Calculate J_s and J_{si}, $1 < j < N$.
4. Extract the sample $x_{\hat{j}}$, $\hat{j} = \min_j J_{sj}$.
5. $L = L + 1$, $S = S \cup \{x_{\hat{j}}\}$
6. If $L < N$ and $J_s <$ max Fitness, the process should turn back to (3), or it should get to (7).
7. S is the extracted feature sample.

In the original KPCA algorithm, the first feature sample is determined through the calculation of the maximum J_s. Middle samples, as the first feature sample, can be applied to get the same results as the original algorithm does. More importantly, middle samples can make the calculation more simplified.

4.3.2.3   Simulation of SKPCA Algorithm

Take some parameters of a rocket engine system as an example. Ten parameters are selected. T refers to the temperature in the combustor, P refers to the pressure of the combustor, P_{ot} is the engine nozzle pressure, F_1 is the flux of the fuel 1, F_2 is the flux of the fuel 2, P_{1r} is the storage pressure of the fuel 1, P_{2r} is the storage pressure of the fuel 2, U_1 is the valve opening of the control valve 1, U_2 is the valve opening of the control valve 2, and P_f is the air supply pressure of control valves. Obtained from data in a certain experiment, the sample set is composed of 50 samples and is to be analyzed through KPCA and SKPCA.

Make the preprocessed data into a $N \times 10$ matrix (N refers to the length of the sample data). After feature extraction and KPCA on the basis of the algorithm mentioned in the last section, the eigenvalues of the matrix can be obtained and are shown in Table 4.1. The table indicates that the accumulative contribution rate of the first four principal components is 97.61%

Table 4.1   The contribution rate of the SKPCA model.

Sequence number	Principal component		
	Eigenvalue of the matrix	Variance contribution rate (%)	Accumulative contribution rate (%)
1	0.4208	0.7864	78.64
2	0.0461	0.0861	87.25
3	0.0362	0.0676	93.01
4	0.0194	0.0362	97.63
5	0.0115	0.0215	99.78
6	0.0009	0.0017	99.95
7	0.0002	0.0003	99.98
8	0.0001	0.0002	100

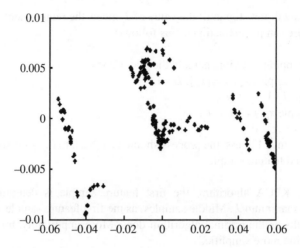

Figure 4.8   The distribution of the first two principal components according to KPCA.

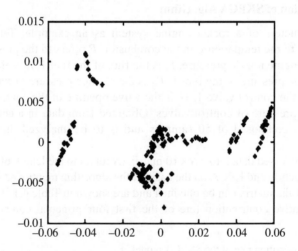

Figure 4.9   The distribution of the first two principal components according to SKPCA.

(>95%), so it can be said that the first four principal components are already able to reflect the most information of the total 10 variables.

The comparison and analysis of simulation results are shown in Figures 4.8, 4.9, 4.10, and 4.11.

Figures 4.9 and 4.11 show that SKPCA has been applied in a certain experiment on some rocket power system. Figures 4.8 and 4.10 show the comparison between SKPCA based on partial samples and KPCA based on all the samples. The results demonstrate that sample feature extraction depending on SKPCA is not simply random reduction of the number of samples, but the basic distributing structure of samples remains unchanged after sample feature extraction. The simulation time of the KPCA model is 0.416 second, while the simulation time

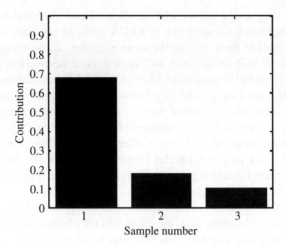

Figure 4.10 KPCA eigenvalue contribution.

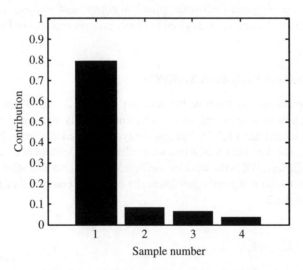

Figure 4.11 SKPCA eigenvalue contribution.

of the SKPCA model is 0.094 second. It is apparent that SKPCA can solve the calculation problem of K and increase the execution efficiency. At the same time, the SKPCA model is basically the same as the principal component model built on all the samples.

4.3.3 Multiscale Fault Detection Based on SKPCA

In practical monitoring process, KPCA has its own demerits. On one hand, process data are often mixed up with noises and interferences, such as white noises and electromagnetic interferences. If such data are directly used for KPCA process monitoring, the results of information processing and analysis will be affected and the confidence level will be lowered. On the other hand, the approach of KPCA needs to calculate the kernel matrix of noise whose size equals to the square of the sample number. If the sample number is large, the calculated amount will be so large that it is quite time-consuming. However, such problem has been solved by the method of SKPCA, which has been discussed in the previous Section 4.3.2. To deal with noises and inferences in data, the method of multiscale KPCA should be adopted. After the orthogonal wavelet transformation for original data, wavelet threshold denoising is supposed to be conducted on the wavelet coefficient at every scale. In the next step, the wavelet coefficient matrix after denoising should be utilized for kernel principal component analysis and wavelet reconstruction. Finally, the multiscale KPCA model is used for online monitoring. In this way, errors can be reduced and the multiscale feature of data can be taken into account.

The number of process variables in the rocket system is very large. If the multiscale analysis goes after the denoising of sample data, the effect will only be theoretically satisfactory. But in practice, it will not work. The reason is that after wavelet denoising, the data have been decomposed and reconstructed, and if the denoised data are analyzed by multiscale KPCA (MSKPCA), they will be denoised and reconstructed for the second time. Therefore, in this section, wavelet denoising and multiscale, principal component analyses are combined, and the method of statistical detection is applied to detect process changes or fault variables.

4.3.3.1 Mathematical Analysis on MSKPCA

MSKPCA has the abilities to improve the accuracy of fault detection due to the multiscale feature of wavelet transformation analysis as well as nonlinearity, dynamic feature, and correlation between data according to KPCA. Specific analyzing procedures are as follows.

Firstly, suppose the data matrix is X ($n \times m$, n refers to the sample number and m refers to the number of variables). WX is the wavelet coefficient matrix obtained after the wavelet transformation of X. $W(n \times m)$ is the orthogonal wavelet operators composed of filter coefficients, shown in Equation 4.12.

$$W = \begin{bmatrix} h_{L,1} & h_{L,2} & \cdot & \cdot & \cdot & & \cdot & \cdot & h_{L,N} \\ g_{L,1} & g_{L,1} & \cdot & \cdot & \cdot & & \cdot & \cdot & g_{L,1} \\ g_{L,1,1} & \cdot & \cdot & g_{L,1,\frac{N}{2}} & 0 & & \cdot & \cdot & 0 \\ 0 & \cdot & \cdot & 0 & g_{L-1,\frac{N}{2}+1} & \cdot & \cdot & & g_{L-1,1} \\ \cdot & \cdot & \cdot & \cdot & \cdot & \cdot & & & \\ \cdot & \cdot & \cdot & \cdot & \cdot & \cdot & & & \\ \cdot & \cdot & \cdot & \cdot & \cdot & \cdot & & & \\ g_{1,1} & g_{1,2} & 0 & \cdot & \cdot & & \cdot & & 0 \\ 0 & 0 & \cdot & \cdot & \cdot & & 0 & g_{1,N-1} & g_{1,N} \end{bmatrix} = \begin{bmatrix} H_L \\ G_L \\ G_{L-1} \\ \cdot \\ \cdot \\ G_m \\ \cdot \\ \cdot \\ \cdot \\ G_1 \end{bmatrix} \quad (4.12)$$

In Equation 4.12, G_m is a $2\log_2^{n-1} \times n$ dimensional matrix composed of wavelet filter coefficients ($m = 1, 2, \ldots, L$, L is the largest number of decomposition layers). H_L is composed of scaling filter coefficients at the maximum layer. The KPCA relation between the two matrixes X and WX can be described by the following two theorems.

Theorem 4.1

The load vectors of X and WX are equal, and the score vector of WX is wavelet transformation of the score vector of X.

Proof: In the data matrix X, if each line's wavelet transformation has the same orthogonal wavelet operator W, Formula 4.13 below is valid.

$$(WX)^T (WX) = X^T W^T W X = X^T X \tag{4.13}$$

This formula proves that the covariance matrix of wavelet coefficients is consistent with that of original data. According to the concept of load vector, 4.13 can further prove that the load vectors of X and WX are equal.

Now that $X = TP^T$, we can get that $WX = (WT)P^T$. Therefore, it is testified that the score vector of WX is wavelet transformation of the score vector of X.

MSKPCA includes KPCA of wavelet coefficients at all layers. Furthermore, as the last step of the algorithm, wavelet coefficients beyond the control limit will be reconstructed by wavelet structure for the reduction of misinformation, and then PCA will be conducted on the reconstructed data again.

The covariance matrix of wavelet coefficients which is consistent with that of original data can be described as the cumulative sum of covariance matrixes at all scales.

$$(WX)^T (WX) = (H_L X)^T (H_L X) + (G_L X)^T (G_L X) + \cdots \\ + (G_m X)^T (G_m X) + \cdots + (G_1 X)^T (G_1 X) \tag{4.14}$$

When exceptional situation occurs for the first time, it will be detected initially by the wavelet coefficients at fine scales. If the exceptional situation continues, it can also be detected by the wavelet coefficients at comparatively coarse scales. As the last safeguard, the scale coefficients (coefficients of low frequency at the coarsest scale) can still detect such abnormity. However, when the exceptional situation returns to normal, the scale coefficients are still beyond the control limit because of their insensitivity to data change, although the wavelet coefficients at fine scales are able to detect the change timely. Accordingly, judging system status by analyzing all layers of wavelet coefficients will cause misinformation or delay.

It is better to make a comprehensive analysis on all scales. The covariance matrix useful for principal component modeling can be calculated through the covariance matrix in abnormal condition.

$$(H_{m-1} X)^T (H_{m-1} X) = (H_m X)^T (H_m X) + \gamma (G_m X)^T (G_m X) \tag{4.15}$$

where $\gamma = \begin{cases} 1, & \text{abnormal results in certain scale of KPCA} \\ 0, & \text{otherwise} \end{cases}$.

The corresponding PCA model can be built on the basis of these matrixes to detect faults in the system. The wavelet coefficients that obtained after wavelet decomposition are mutually independent with each other. It means that wavelet coefficient sequences do not have the feature of autocorrelation. Thus, wavelet coefficient modeling can well solve the problem of sequence correlation in traditional KPCA modeling. Wavelet transformation belongs to orthogonal transformation, so the KPCA model at each scale will not change the correlation between variables.

4.3.3.2 Algorithm Analysis on MSKPCA

Due to impacts of various events with different local time-frequency characteristics, MSKPCA usually has complicated multiscale features, but traditional statistical and monitoring approaches, such as PCA, are only on a single scale. MSKPCA belongs to a multiscale monitoring approach in which information on multiscales can be utilized simultaneously. The fault-detecting frame built on the basis of MSKPCA is shown in Figure 4.12. In consideration of a data matrix $X \in R^{N \times m}$ in a process database, the first step is to conduct wavelet transformation to decompose all variables on different scales. Then the approximate part A_J and the detail part $D_1, D_2, \ldots, D_J$ of the data base are obtained (suppose the biggest decomposition scale is J). After reconfiguration of wavelet coefficients on all scales, the reconstructed $J+1$ dimensional matrix $X^{[0]}, X^{[1]}, \ldots, X^{[J]}$ is obtained. From Formula 4.14, it can be known that $X = X^{[0]} + X^{[1]} + \cdots + X^{[J]}$. According to MSKPCA, the augmented matrix is $\tilde{x} = [(x^{[0]})^T (x^{[1]})^T \cdots (x^{[J]})^T]^T$, so the dot product kernel function in the nonlinear feature space is defined in Formula 4.16.

$$k\left(\tilde{x}_s, \tilde{x}_t\right) = \exp\left(-\frac{\left\|\tilde{x}_s - \tilde{x}_t\right\|^2}{c}\right)$$

$$= \prod_{j=0}^{J} \exp\left(-\frac{\left\|\tilde{x}_s^{[j]} - \tilde{x}_t^{[j]}\right\|^2}{c}\right) \qquad (4.16)$$

$$= \prod_{j=0}^{J} k\left(\tilde{x}_s^{[j]}, \tilde{x}_t^{[j]}\right)$$

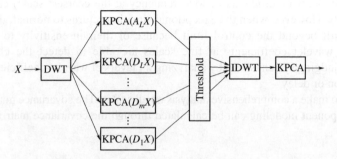

Figure 4.12 Analytical procedures of MSKPCA.

In Formula 4.11, the third equation meets the requirements of Gaussian kernel function $\left(k\left(\tilde{x}_s^{[j]}, \tilde{x}_t^{[j]}\right)\right.$ represents the local kernel function on the scale of J). In MSKPCA models, all local kernel functions are allowed to be inconsistent for the sake of flexibility of applications. To avoid complicated calculation, which will affect algorithm efficiency, all local kernel functions should be all consistent. Then make KPCA analysis on the reconstructed matrix and conduct dimensionality reduction for extraction of main information, which can be used for fault detection.

MSKPCA can realize nonlinear transformation of multiscaled data on the basis of kernel functions, and make PCA analysis on these data in a linear space. The establishment procedures of MSKPCA models are as follows.

1. Choose 2^N (N should be positive integer) normal data samples and conduct multiscaled wavelet decomposition on each line of these samples through wavelet threshold denoising mentioned in Chapter 2. After getting the wavelet coefficients at each fine scale and each coarse scale of every line, the wavelet coefficient matrix is supposed to be formed at each scale.
2. Make KPCA analysis at each scale and calculate the covariance matrix and principal component values. Then choose the suitable number of principal components and calculate SPE statistics as well as the control limit. Make the SPE control limit as the threshold and select the wavelet coefficients that are greater than or equal to the threshold.
3. Make a suitable combination of scales of significant events and reconstruct the selected values and threshold values at these scales. Make PCA analysis on the reconstructed matrix and calculate T^2 and SPE statistics as well as the control limit. Confirm the number of principal components and build integrated principal component models through which fault detection can be conducted.

For the data collected in practical systems, faults may occur in different frequency range, which proves that it is quite necessary to conduct multiscaled fault diagnosis. Take the noise signal in Figure 4.13 which is featured by various kinds of faults, such as constant deviation, slowly varying characteristic, mutation, and high-frequency sine. High-frequency sine are taken as examples and are shown in Figure 4.13.

Make sym wavelet decomposition on the signal in Figure 4.13, and the detailed features at each scale after decomposition are shown in Figure 4.14. It is easy to be seen in Figure 4.13 that the detailed signals on the first layer describe high-frequency sine and mutation faults in the composite signal. The detailed signals on the second layer can describe all four types of faults, but only high-frequency sine faults are clearly identified. The detailed signals on the third layer mainly reflect constant deviation and slowing varying faults, those on the fourth layer clearly display slowly varying faults, and the ones on the fifth layer clearly show constant deviation faults. After analyzing the detailed features in Figure 4.14, we can draw a conclusion. Constant deviation and slowly varying faults in practical systems are mainly described by coarse-scaled detailed features. While mutation and high-frequency sine faults in practical systems are mainly described by fine-scaled detailed features.

4.3.4   Improved Principles of Multiscale Kernel PCA

Although traditional MSKPCA approaches have the superiority of multiscaled modeling, they still have two shortcomings. Firstly, most operating data contain noises and need denoising. However, it is inevitable to increase the amount of calculations if reducing

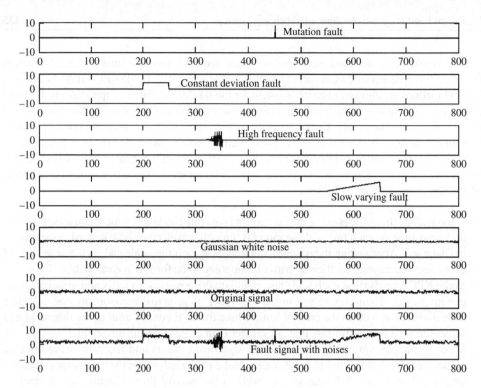

Figure 4.13 Noisy fitting fault signal.

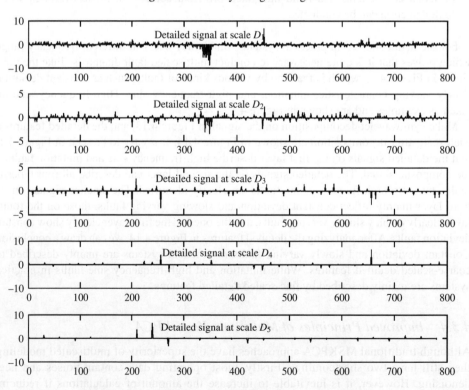

Figure 4.14 Discrete wavelet decomposition of fitting signals.

noises initially and then making MSKPCA analysis on these data. Therefore, these methods should not be effectively used for fault detection and diagnosis in the industry. Secondly, MSKPCA unanimously chooses SPE or T^2 control limit as the threshold value to eliminate noises in every variable. Such noise reduction method is realized through ignorance of some information of fine-scaled models. In practical situations, the noise level of each variable is different, so a unanimous standard is not appropriate for noise reduction. Moreover, information at each fine scale usually contains information of process changes, which is supposed to be used to assist fault detection rather than being simply ignored.

Due to these two shortcomings of MSKPCA, an improved MSKPCA approach is adopted. First, the standard noise deviation δ and the threshold value T of each scale coefficient are calculated after the orthogonal wavelet decomposition of each line in sampled data. Wavelet threshold is adopted to denoise eliminate noises and abnormal points in each scale coefficient matrix. Afterward, KPCA and multiscaled analysis are utilized to analyze low-frequency coefficients and after-denoising high-frequency coefficients, and wavelet transform is used to reconstruct wavelet coefficients beyond the control limit. Furthermore, KPCA is used to analyze the reconstructed data again to meet the goal of combining wavelet denoising and MSKPCA. Such combination can not only avoid the impact of different levels of noises on scale modeling but also optimize the algorithm structure and improve executive efficiency, which can help accomplish fault detection more effectively.

The improved principle of MSKPCA is shown in Figure 4.15, where WTDN represents wavelet threshold denoising.

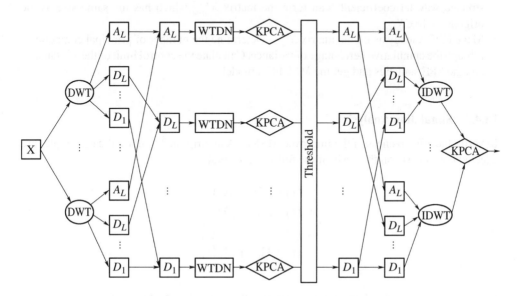

Figure 4.15   The principle diagram of MSKPCA.

4.3.4.1   Improved MSKPCA Fault Detection Algorithm

Improved MSKPCA algorithm can eliminate noises in process data by using wavelet threshold denoising. Meanwhile, it can combine with MSKPCA as a more simplified algorithm structure to increase fault detection efficiency. The algorithm is as follows.

1. Select data samples according to the algorithm of feature sample extraction mentioned in Section 4.3.2, and from a new sample set to increase the efficiency in the condition that the SKPCA model is basically the same as the principal component model built by all the samples.
2. Select 2^N (N is positive integer) normal data samples for wavelet decomposition. Each line of data represents a corresponding variable, so there forms the original data matrix $X_{n \times m}$ (n is the number of samples and m is the number of variables). Centralize and standardize each line in the matrix to make the base point of each variable basically the same.
3. Conduct orthogonal wavelet transformation on each data line in the standardized data matrix $X_{n \times m}$, then we can get wavelet coefficients on every coarse scale and fine scale. Respectively calculate standard noise deviation δ and threshold value T of each line of each scale coefficient, preparing for wavelet threshold denoising.
4. Combine the coefficients on the same scales as a coefficient matrix. Each matrix stands for the change trend of a scale. Make PCA analysis on the wavelet coefficient matrix of each layer and calculate the covariance matrix and the principal component value. Then, select the suitable number of principal components and calculate the T^2 statistics, the SPE statistics, and their respective control limit, which constitutes the principal component model on all scales.
5. Whether the T^2 statistics and the SPE statistics run beyond the standard index can reflect if there contains important information on each scale. On the scale where there is important information, the control limit of the SPE statistics is seen as the threshold value, and each line of data is to be operated of threshold denoising and be reconstructed. The reconstructed wavelet coefficients can form the matrix $X'_{n \times m}$ which has the same size as the original matrix $X_{n \times m}$.
6. Make KPCA analysis on the matrix $X'_{n \times m}$. Determine the number of principal components through the cumulative percentage of variance. Calculate the control limit of the T^2 statistics and the SPE statistics and get the MSKPCA model.

4.3.4.2   Simulation Analysis

To verify the effectiveness of the improved MSKPCA algorithm, 256 normal data samples are selected for analysis on the basis of the following formulas.

$$\begin{cases} \tilde{x}_1(t) = 1.5 * N(0,1) \\ \tilde{x}_2(t) = 2.5 * N(0,1) \\ \tilde{x}_3(t) = \tilde{x}_1(t) + 2\tilde{x}_2(t) \\ \tilde{x}_4(t) = 3\tilde{x}_1(t) - \tilde{x}_2(t) \end{cases} \tag{4.17}$$

These four formulas above constitute the matrix $\tilde{X}(k) = [\tilde{x}_1(k) \quad \tilde{x}_2(k) \quad \tilde{x}_3(k) \quad \tilde{x}_4(k)]$.

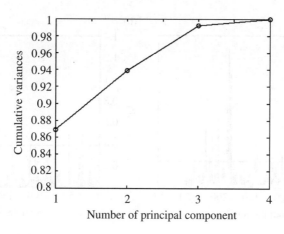

Figure 4.16 The cumulative explain degree of principal component models with different number of principal components.

Noises in the data are eliminated by the approach of building improved MSKPCA model, and a comprehensive KPCA model is established. The number of principal components is obtained through the cumulative percentage of variance. Figure 4.16 shows the cumulative explain rate of principal component models with a different number of principal components.

Figure 4.16 shows that the first three principal components can explain 99% data change, so they are preserved as the principal components in the comprehensive principal component model. The control limit of credibility is 0.3832 and the SPE statistics is 95%; the control limit of credibility is 6.8542 and the T^2 statistics with 95%. The wavelet denoising threshold value λ is 3.5482, which is calculated according to the rule of Sqtwolog. In this way, the improved MSKPCA model is obtained. Meanwhile, the traditional KPCA and MSKPCA principal component models are also being established respectively for approach comparison.

After the model is built, the mean shift fault whose amplitude is 0.5 is introduced between the 176th and the 225th sample points of each variable, and abnormal points are also introduced into the 20th, the 50th, and the 120th sample points of the first variance. KPCA, MSKPCA, and improved MSKPCA are used to make fault detection of data. The following are T^2 and SPE monitoring figures as well as the performance comparison tables of each approach.

Figure 4.17a and b shows that KPCA can detect partial faults, but there exist some misinformation and failures. For example, in the T^2 monitoring figure, the faults between 185 and 200 are not detected, and the introduced abnormal points as well as the 100th and the 145th sample points are taken as faults mistakenly. The SPE monitoring figure has less misinformation, but its abnormal points are also taken as faults. Figure 4.17c and d shows that MSKPCA has little misinformation. Only the sample points from 190 to 196 in the T^2 monitoring figure are detected faultless, but in the two monitoring figures, the abnormal data of the 20th, the 50th and the 120th sample points are taken as faults. Figure 4.17e and f shows that the improved MSKPCA can not only detect faults precisely but also eliminate abnormal points effectively with only very few failures. It can also be seen that the fault has been detected just at the 176th sample point by the improved MSKPCA, while the other two approaches are slower to detect the fault.

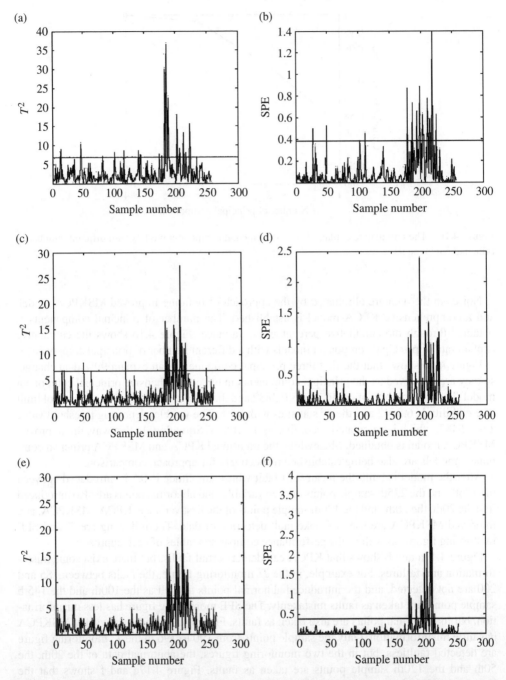

Figure 4.17 T^2 and SPE monitoring diagrams. (a) and (b) donate T^2 monitoring and SPE monitoring by KPCA, (c) and (d) donate T^2 monitoring and SPE monitoring by MSKPCA, (e) and (f) donate T^2 monitoring and SPE monitoring by improved MSKPC.

Table 4.2 Performance comparison between different PCA approaches on the basis of SPE statistics.

Algorithm	Performance index (%)	
	Rate of erroneous detection	Rate of missing report
KPCA	13.8	26.7
MSKPCA	7.13	12.9
Improved MSKPCA	0	2.03

Table 4.3 Time of different PCA approaches.

Method	KPCA	MSKPCA	Improved MSKPCA
Time/seconds (T^2)	10.23	21.36	19.78
Time/seconds (SPE)	10.47	21.58	20.09

Table 4.2 provides the rates of erroneous detection and missing report of different fault detecting approaches on the basis of SPE statistics. For the improved MSKPCA, the rate of erroneous detection is zero, which means that noises and abnormal points are apparently eliminated by using wavelet threshold denoising. The rate of missing detection of the improved MSKPCA is 2.03%, a great decrease compared with that of traditional KPCA and MSKPCA.

Table 4.3 shows that improved MSKPCA is faster than traditional MSKPCA, and about 6–7% of time is saved on average.

To sum up, improved MSKPCA is obviously superior to other two approaches because it takes less time and promotes the detecting quality.

In practice, many data contain a large amount of noises which are mixed up with fault information. This leads to erroneous detection by using traditional analysis approaches. Due to the large number of testing variables and processing data, the sample feature extraction followed by MSKPCA analysis is applied to promote the efficiency of KPCA analysis. The basic idea of improved MSKPCA is that wavelet threshold denoising is conducted for wavelet coefficients of all layers after the orthogonal wavelet transformation of original data. And then wavelet coefficient matrixes are analyzed on all scale by KPCA. The wavelet coefficients are reconstructed with the thresholds that are control limits of T^2 and SPE statistics. At last, the multiscaled data matrix is obtained and the comprehensive PCA model is constructed.

The improved MSKPCA promotes the efficiency of the KPCA algorithm through sample feature extraction and it also combines wavelet denoising with MSKPCA. It can be applied to eliminate noises in original data as well as reducing the rates of erroneous detection and missing detection on different scales. In this way, time can be saved and the accuracy of fault diagnosis can be raised. Test results show that improved MSKPCA can be used to detect fault information faster and more accurately.

4.3.5 Online Fault Detection Based on Sliding Window MSKPCA

During offline data analysis, some representative data are selected to build the mapping relation between input data and output data. In general, the model will not change any more after its establishment. When it is applied into a time-varying system, the system is only related to data around operating points instead of data far away. After a period of operation, the system's operating area will change. The model, if built with fixed data, will not be capable to describe practical situation of the system with the time and the conditions changing.

Accordingly, sliding window is supposed to be adopted so that new data are added continuously and the monitoring model can be updated automatically. It guarantees the accuracy of the model as well as the accuracy rates of fault detection.

In desired situation, the operating parameters of the engine are normally stable, which means that the mean value and the variance stay unchanged. Such system is regarded as a time-invariant system which has no relation to time. However, the operating parameters of the engine will gradually change over the time because of abrasion, burn-in, raw material change, or sensor failures. It will negatively influence the accuracy of the model to emerge in a gradual way, so it is not easy to detect them.

In conclusion, a time-invariant MSKPCA model that is used to monitor a time-varying system may cause errors in fault detection and diagnosis and lead to misinformation as time goes on. Therefore, it is urgent to build a self-adaptive model and update the monitoring model automatically through a collection of normal real-time data. Sliding window is more effective if it is combined with MSKPCA. The data of real-time collection can make the monitoring model update automatically. If the KPCA monitoring model can adapt to normal parameter drifts in the time-varying system, the speed and the accuracy rate of fault detection will be definitely improved.

4.3.5.1 The Basic Idea of Sliding Window

Practical and normal sample data of each engine are collected constantly to update old sample data which is used for original modeling. That means the number of samples in the new sample set remains the same. The newly formed sample set is applied for modeling, determining the number of principal components, and calculating the statistics and the control limit. The updated KPCA model is adopted for more effective fault detection.

Suppose the length of the sliding window is w and the step length is h. The sliding window is

$$X_{w+h} = [x_{h+1}, \ldots, x_m, \ldots, x_{w+h}] \tag{4.18}$$

Figure 4.18 displays the sliding window.

The length of the sliding window should not be too short, or it is impossible to form a covariance matrix. However, the length of the sliding window should not be too long either, because the dimensionality of the kernel matrix K will be high and the calculated amount will

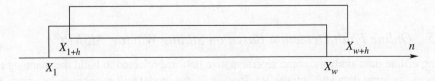

Figure 4.18 The sliding window.

be increased to a large extent. Therefore, it is important to select a reasonable window. Furthermore, the selection of the step length also depends on the concrete situation of research subjects. If system parameters drift too fast, a small value of corresponding step length should be adopted. In some special situation, the step length can be 1, which means the KPCA model can be updated as long as a normal data is collected. However, the problem is that the over-fast updating frequency will result in increasing calculated amount, which is not beneficial to online monitoring. For a system with slow parameter drifts, it is not necessary to update the KPCA model too frequently and the step length can be long.

4.3.5.2 Online Fault Detection Algorithm Based on Sliding Window MSKPCA

Algorithm procedures are as follows.

1. Select normal sample data for initializing the sliding window. The length of the sliding window is $w = 2^n$ (n is positive integer) for the sake of wavelet transformation. The step length is h.
2. Calculate the mean value and the variance of window data. Standardize the sliding window data with the mean value and the variance, thus the base point of each variable will be the same.
3. Make orthogonal wavelet transformation on each line of the standardized data matrix, and make KPCA analysis after wavelet threshold denoising of all layers of wavelet coefficients. Calculate the covariance matrix and the principal component score value of wavelet coefficients. Select a reasonable number of principal components according to the method of cumulative variance percentage. Calculate the T^2 statistics and the SPE statistics as well as their respective control limit, getting the principal component model on all scales.
4. Make a combination of the scales of significant events detected. Reconstruct the wavelet coefficients on these scales. Calculate control limits of the T^2 statistics and the SPE statistics, getting the MSKPCA model.
5. Collect new data x_{new} and standardize the data with the mean value and the variance obtained through step 2. Repeat step 3, and compare with the control limits of models on all scales in step 3. If beyond the control limit, there is abnormal situation on the scale, so it is urgent to detect faults on the ultimate scale. Get multiscaled data according to step 4 and make KPCA analysis. Calculate the T^2 statistics and the SPE statistics and judge whether the statistics are beyond the control limit on the basis of the model in step 4. If the T^2 statistics and the SPE statistics are not beyond the control limit, the newly collected sample x_{new} is regarded as normal, and make accumulation operation $i = i + 1$. Otherwise, the sample is classified as fault and accumulation operation canceled.
6. If h continuous times of new data collection are normal ($i = h$), the data window will be updated, and the window will move six steps forward. The practical testing data from h continuous times of collection are added into normal samples. Meanwhile, to make the window length unchanged, h old samples need to be deleted from w normal samples in the original window. The normal sample set is updated in such a way. Then suppose $i = 0$, and repeat the steps 2, 3, and 4. If $i < h$, the window does not move, and the normal sample set remains the same without any updates. Repeat step 5 and continue to detect faults with the original model.
7. Draw a contribution figure about the SPE statistics and selected principal component of each variable according to the faulted data sample in step 5. Determine the variables that cause faults and failures.

4.3.6   Examples

In this section, a vibration signal in practical testing functions as a diagnosis subject is discussed. The effectiveness of multiscaled fault diagnosis is verified through data experiments on this signal. After getting 8×2000 data samples (as shown in Figure 4.19, the faults occur in two sample sections—619–632 and 938–984 which are the locations of sensor No. 1 and sensor No. 2), data experiments are conducted by the approaches of traditional PCA, traditional KPCA, SKPCA, MSKPCA, and MW-MSKPCA, and further analysis are made on these experimental results.

Algorithms of the five approaches have been elaborated in detail in the previous sections, so in this section, how to make fault diagnosis by the approaches of MSKPCA and MW-MSKPCA is introduced. The monitoring performance indexes and parameters of other algorithms are the same as those of the MSKPCA algorithm.

8×2000 normal data are selected as training samples. Wavelet decomposition is made with an edge correction filter in the data window in a length of $N = 8$ (the decomposition scale is $L = \log_2^{2000} - 5 = 5.956 \approx 6$). After getting wavelet decomposition coefficients of all layers, threshold denoising is conducted for these coefficients by the formula $T = \sigma \sqrt{2 \ln(n)}$, namely

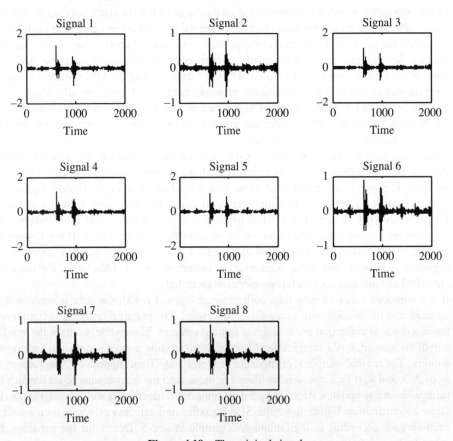

Figure 4.19   The original signal.

$T = (0.002753/0.6745) \cdot \sqrt{2\ln(2000)} = 0.01591$. Then self-adaptive PCA analysis is made on the wavelet coefficients after denoising on each scale. The original data block of RPCA is set as 8×256, and then it ought to calculate corresponding wavelet coefficient covariance matrix, principal component score vector, and load vector. After that, the amount of principal components is chosen according to the principal component selection rule. The value of statistic control limit is worked out as 95% according to the Hotelling T^2 statistics and the definition of SPE. After the derivation of the reference principal component model, the amount of 8×2000 faulted data is input to the model as testing samples. If there is any malfunction accruing in the system, the relevant significant events can be detected on each scale. The wavelet coefficients are reserved, which are beyond the statistic control limit on each scale. These wavelet coefficients are reconstructed to get a reconstructed data matrix. It is to work out the wavelet coefficient covariance matrix, principal component score vector, and load vector of the reconstructed data matrix. T^2 and Q statistics of the reconstructed data matrix are calculated according to the Hotelling T^2 statistics and the definition of SPE. If the T^2 and Q statistics are beyond the statistic control limit of the reference principal component model, there will be a fault warning, and fault diagnosis is realized.

The following are SPE and T^2 figures for eight vibration signals obtained after the diagnosis through traditional PCA, traditional KPCA, SKPCA, MSKPCA, and MW-MSKPCA.

Figure 4.20a–e shows the SPE statistic monitoring for the algorithms of traditional PCA, traditional KPCA, SKPCA, MSKPCA, and MW-MSKPCA, respectively.

In Figure 4.20a, there is no fringe effect because the issue of filtering is not involved. However, there still occurs erroneous detection and missing report. The faults during 624–628 and 954–970 are missed. It is illustrated in Figure 4.20b, c, and d that although faults can be detected through the approach of SKPCA, the problem of missing report still happens near the 986th sample point. The MSKPCA algorithm works well midway through the whole statistic process and the location of faults can be diagnosed accurately. However, there is erroneous detection near the 1900th sample point, which shows the negative impact of edge effect. Compared with traditional KPCA and SKPCA, the algorithms of MSKPCA and MW-MSKPCA diagnose more accurately without any erroneous detection. Furthermore, compared with the traditional algorithm of KCPA, the algorithms of SKPCA, MSKPCA, and MW-MSKPCA, do better in noise elimination. However, the improved MSPCA algorithm performs not well in terms of noises eliminating on sample points in edge position. By contrast, fault signals are monitored more clearly throughout the whole process by the algorithms of MSKPCA and MW-MSKPCA.

From Figure 4.20a–e, we can draw a conclusion that the algorithm of threshold denoising PCA is more advantageous than the PCA algorithm, which does not take noises into account. In threshold denoising, the algorithm with an edge filter is better than the PCA algorithm which eliminates noises in signals directly.

After the completion of SPE charts, the diagnosing effect of the five methods is analyzed in this section from the perspective of T^2 statistical. The T^2 statistical monitoring charts acquired are shown in Figure 4.22. The conventional PCA, traditional KPCA, SKPCA, MSKPCA, and MW-MSKPCA algorithm are adopted in Figure 4.21a–e

Compared with Figure 4.20, Figure 4.21 tends to be more complicated. This is due to the features of the two statistical approaches: SPE statistics reflect the sum of squared error between the measurement value and the pivot model predicted value. While T^2 statistics reflect changes of the internal pivot vector model of the pivot model. In case of the abnormalities such as the fluctuations

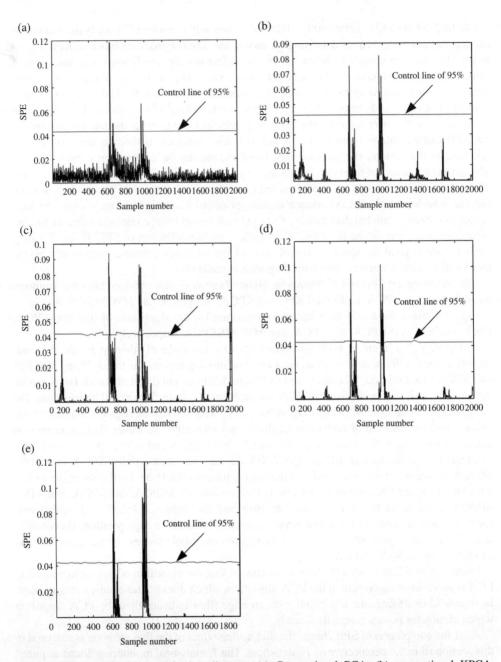

Figure 4.20 SPE statistic monitoring diagram. (a) Conventional PCA, (b) conventional KPCA, (c) SKPCA, (d) MSKPCA, and (e) MW-MSKPCA.

of common variable mean value and amplitude as well as sensor failure, the average value of T^2 and SPE statistics will change correspondingly. In the actual process of monitoring, it is necessary to analyze T^2 and SPE statistical variation in an integrated way rather than relate the changes of process conditions, process failure, or sensor failure with certain statistics separately.

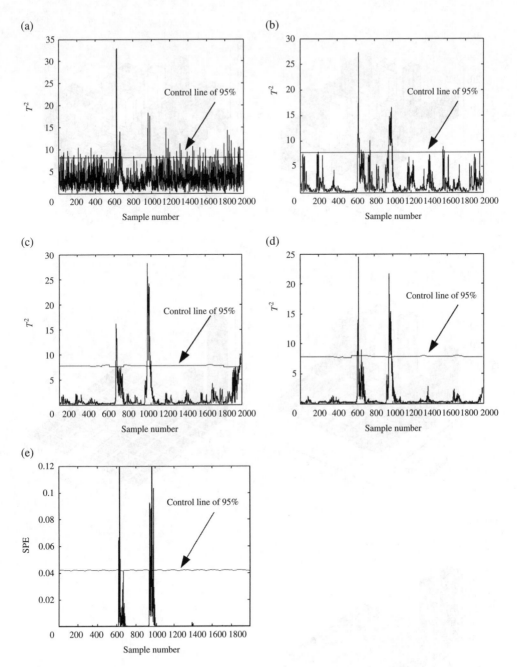

Figure 4.21 T^2 statistics of empirical data. (a) Conventional PCA, (b) conventional KPCA, (c) SKPCA, (d) MSKPCA, and (e) MW-MSKPCA.

From Figure 4.21a, the traditional PCA algorithm leads to errors in diagnostic result in various degrees because of ignorance of the noise disturbances. T^2 statistics reflect the changes of internal pivot vector of the vector model, while the noise effects are cumulated in the pivot space through the PCA algorithm without noise-removing measures. The SPE statistics reflect

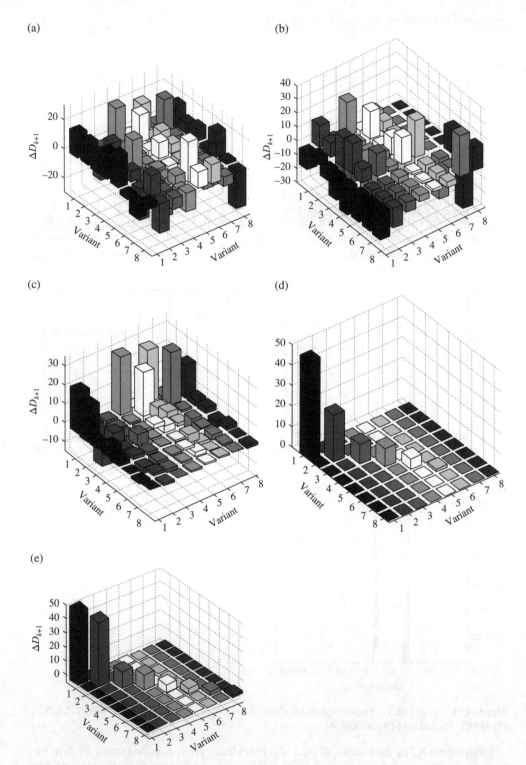

Figure 4.22 Tridimensional contribution of empirical data. (a) Traditional PCA, (b) traditional KPCA, (c) SKPCA, (d) MSKPCA, and (e) MW-MSKPCA.

the sum of squared errors of the measurement value and pivot model predicted value, by which noise effects are offset to some degrees. It is illustrated that SPE statistics monitoring produces clearer results than the T^2 monitoring chart.

From the troubleshot monitoring opting results of Figure 4.21b–e, we can see that the conventional MSPCA algorithm contains a lot of noises in fault diagnosis, and obvious faults have occurred at 630th and 1580th sampling point. The improved MSPCA algorithm in Figure 4.21c shows clearer fault signal of the whole monitoring process, but there are still faults at the 1900th sampling point. Compared with the former two situations, there are fault reports at 689th sampling point in Figure 4.21d while it is more serious in Figure 4.21e including errors in the sample interval from 692th to 694th and omitted errors in the sample interval from 961th to 964th. Compared with the three traditional algorithms above, MSKPCA and MW-MSKPCA are much clearer and simpler in diagnosing.

After the time diagnosis of equipment faults through multiple statistics monitoring charts, the location diagnosis will be carried out through tridimensional contribution charts. The diagnosis results are shown in Figure 4.22 and the tridimensional contribution chart is acquired through algorithms in Figure 4.22a–e which adopt traditional PCA, traditional KPCA, SKPCA, MSKPCA, and MW-MSKPCA, respectively. From Figure 4.22a–c, we can see that the data collected by the eight sensors of the device is in a mess, making it difficult to determine the accurate positions of the fault occurred. From Figure 4.22d–e, it is clear that the failure occurred between the first and second sensor, and the second sensor in Figure 4.22e contributes more than Figure 4.22d, showing that the failure occurs in the second sensor. So we come to this conclusion: in terms of fault location, MSKPCA and MW-MSKPCA have similar effects, but the two are obviously better than MSPCA and improved SKPCA algorithm.

In order to describe the accuracy of the five algorithms in terms of fault diagnosis from the quantitative perspective, we define fault diagnosis accuracy A:

$$A = \sqrt{\frac{\sigma_{T^2}^2 + \sigma_{SPE}^2}{2}} \tag{4.19}$$

where, σ_{T^2} and σ_{SPE} represent the fault diagnosis accuracy of T^2 and SPE statistics monitoring charts, that is,

$$\begin{cases} \sigma_{T^2} = 1 - \left(\eta_{f.T^2} + \eta_{o.T^2} \right) \\ \sigma_{SPE} = 1 - \left(\eta_{f.SPE} + \eta_{o.SPE} \right) \end{cases} \tag{4.20}$$

where

$$\begin{cases} \eta_f = \dfrac{n}{N} \\ \eta_o = \dfrac{m}{N} \end{cases} \tag{4.21}$$

In which, η_f is the false alarm rate, η_0 is the missing alarm rate, n is the number of false alarm samples, m is the number of missing alarm samples, and N is the total number of fault samples.

Table 4.4 Comparison of algorithm accuracy.

Algorithm	Statistics	Missing alarm rate (%)	False alarm rate (%)	Accuracy rate (%)
Traditional PCA	T^2	6.56	40.98	53.28
	SPE	36.07	9.84	
Traditional KPCA	T^2	3.28	13.11	84.43
	SPE	6.56	8.20	
SKPCA	T^2	1.64	11.48	89.43
	SPE	3.28	4.92	
MSKPCA	T^2	1.64	4.92	94.26
	SPE	3.28	1.64	
MW-MSKPCA	T^2	4.92	6.56	93.57

According to formula (4.18), (4.19), and (4.20), we obtain the accuracy rates of the aforementioned five algorithms, the results of which are shown in Table 4.4.

From Table 4.4, it is clear that traditional PCA algorithm has the lowest accuracy rate of 53.28%, and the MSKPCA algorithm is highest at 94.26% followed by MW-MSKPCA with an accuracy rate of 93.57%. The following is the comparative analysis:

1. PCA is an algorithm based on data covariance structure. Under PCM algorithm, the pivot model will no longer change once established, but the experimental data is time-varying. Therefore, PCA is not accurate for descriptions of dynamic properties of nonstationary processes.
2. PCA works on a single scale rather than on a multiscale, so it cannot fully extract information contained in dynamic data.
3. PCA is a linear transformation in nature and has a deficiency in dealing with the nonlinear problem. Therefore, it is normal that the algorithm shows low accuracy in the handling of the dynamic nonlinear nonstationary data from the device.
4. As MSPCA considers the multiscale characteristics of data, it is more accurate than single-scaled PCA and KPCA mode. However, the traditional MSPCA still belongs to a fixed model data-driven algorithm. The fixed model cannot reflect the time-varying process of data, and it ignores the effects of noise on diagnosis results, so its fault diagnosis accuracy rate stays only at 84.43%, lower than the improved MSKPCA.
5. Both MSPCA and MSKPCA algorithm consider the multiscaling and time-varying characteristics of data as well as a deviation of the statistic's model caused by noises, but MSPCA algorithm just eliminates noises of threshold value of the small wavelet coefficients directly while ignoring the edge effects of the process of noise elimination. This can be found in 1900th sampling point in Figures 4.20d and 4.21d that show obvious false alarm.
6. MW-MSKPCA has lower accuracy than MSKPCA, mainly because the edge correction filter is better than the signal symmetric extension, but the edge correction filter is more complicated in design.

4.4 Fault Tree Analysis Based on Ant Colony Algorithm

In order to accurately and effectively describe the relationship between system state conditions and the faults, the system hierarchy analysis is adopted to divide and decompose the system, set up, and analyze the state tree that describes the system status, and integrate with the fault chart to form the conditional fault graph for autonomous fault description and analysis. Based on this, the ant colony algorithm is introduced to determine the optimal test sequence of fault chart and guide system to make a strategic decision of multi-faults status. Apply them to the launch vehicle's control system, and a characteristic example is offered here. The conditional fault graph can effectively describe the influence of status and conditions on the fault while the ant colony algorithm could carry out dynamic path selection in a real-time and adaptive manner.

4.4.1   Layer Structural Analysis of System

There are a number of parallel subsystems in the rocket system, which could be divided into several logic levels. As the system has multiple layers in both structure and functions and the system and subsystem are closely coupled, it is very difficult to analyze the system as a whole. Therefore, the object-oriented problem reduction method is adopted for problem solution, which decomposes the whole diagnosis problem into several sub-problems by certain strategy.

4.4.1.1   Mathematical Hierarchical Decomposition Model Based on Structure and Behavior

A complex device consists of interconnected subsystems in a certain way and has specific features and characteristics. Therefore, the complex system can be modeled as a hierarchical collection of the hierarchy digraph.

Set the control system as S, and after hierarchical decomposition of S according to its structure and function, the S could be expressed as a collection of levels:

$$S = \{S_1, S_2, ..., S_i, ..., S_n\}, \quad i = 1, 2, ..., n \tag{4.22}$$

After hierarchical decomposition of S, the hierarchical chart of one level D could be defined as:

$$D = (V, E) \tag{4.23}$$

In which, $V = \{S_1, S_2, ..., S_n\}$ is the node collection, $E = S \times S$ is the edge collection that connects the nodes, composed of connection R between subsystems S_i.

Connection relationship R is defined based on S, in which $R \subseteq S \times S$ and $S_i R S_j$, S_i, $S_j \in S$. This connection constitutes the whole paths for fault spreading in the control system. The node-set V of hierarchy digraph D consists of the monitorable node-set V_M and non-monitorable node-set V_N, and meets the requirements:

$$\begin{cases} V = V_M \cup V_N \\ V_M \cap V_N = \Phi \end{cases} \tag{4.24}$$

In D, the directed edge from S_j to S_k is represented by $e_{jk} = (S_j, S_k)$, and

$$e_{jk} \in E \quad \text{and} \quad S_i R S_j \tag{4.25}$$

The connectivity matrix A of hierarchy digraph D is represented by:

$$A = \{a_{jk}\} \tag{4.26}$$

where $a_{jk} = \begin{cases} 1, & e_{jk} \in E \\ 0, & \text{otherwise} \end{cases}$.

The connectivity matrix A could be used to calculate the availability matrix of system S:

$$M = (m_{jk}) = [I + A]^k = [I + A]^{k+1} \tag{4.27}$$

where $k \geq k_0$, k_0 is a positive integer and I is the unit matrix. The generation node set and descendant node set are two functions of node set V in hierarchical directed digraph D, which is defined as:

$$A_M(S_j) = \{S_k | m_{jk} \neq 0, m_{jk} \in M, S_j, S_k \in V\}$$
$$V_i = [A_M(S_j) - V_1 \cdots V_{k-1}], \quad k = 2, \ldots, m \tag{4.28}$$

Through the availability matrix M, the node set V in hierarchical directed digraph D could be decomposed into m hierarchies, namely $V_1, V_2, \ldots, V_m$, in which,

$$V_i = \{S_j | D_M \cap A_M(S_j) = A_M(S_j)\} \tag{4.29}$$

$$V_i = [A_M(S_j) - V_1 \cdots V_{k-1}], \quad k = 2, \ldots, m \tag{4.30}$$

Here, $m (\leq n)$ is a positive integer that meets $V - V_1 - \cdots V_m = \Phi$. This is called hierarchical structure. This object class meets the following conditions:

1. $\bigcup_{i=1}^{m} = V$.
2. $V_i \cap V_k = \Phi, j \neq k$.
3. For $S_j, S_k \in V$ and one of the following conditions is met: if $m_{jk} = m_{kj} = 1$, then S_j and S_k are in the same circuit; if $m_{jk} = m_{kj} = 0$, then S_j and S_k are not connected.
4. Outside the nodes of object V_j, the edges can only be found in the nodes of object class V_k.

The decomposition starts from the system level and expands to the subsystems, sub-subsystems, components and parts in a step by step manner. S, a complex equipment with single node, constitutes itself the first level of hierarchical decomposition while hierarchy digraph constitutes the second-tier of control system hierarchical decomposition model. On this basis, according to the hierarchical decomposition model, the system continues hierarchical decomposition of the structure and functions of each subsystem. The hierarchical decomposition model is defined as:

$$D(V_j, E_j), \quad j = 1, \ldots, n \tag{4.31}$$

Collection $\{D_j\}$, composed of the n-layered hierarchy digraph, forms the tertiary level of the hierarchical decomposition model. Just like this, the hierarchical structure correspondent to node V connecting hierarchy i and $i-1$ of the hierarchical decomposition model is expressed as:

$$D_{ijk} = \left(V_{ijk}, B_{ijk} \right), \quad j = 1, \ldots, n \qquad (4.32)$$

In conclusion, the mathematical model of hierarchical decomposition of the control system S according to structure and function could be expressed as:

$$H = \left\{ L_i \right\}, \quad i = 1, 2, \ldots, l \qquad (4.33)$$

where l means the level of the hierarchical decomposition model, L_i represents level i of the hierarchical decomposition model, and

$$L_i = \left\{ D_{ijk} \right\}, \quad j = 1, \ldots, n; \; k = 1, \ldots, m_j \qquad (4.34)$$

With the hierarchical decomposition model above, decompose the control system S by structure and function. The top layer is control system, the second level is the subsystem S_j that constitutes the control system S, and the third level is the sub-subsystem S_{jk} that forms subsystem S_j, ...; level i represents the components correspondent to the sub-subsystem on level $i-1$. Decompose the structure until the required level is acquired.

4.4.1.2 Establishment of System Conditional Fault Graph Based on Hierarchical Decomposition

In order to describe the information in a complete and standard way, this section proposes a description model of conditional fault graph in order to facilitate the automated descriptions and automated analysis using computer.

Description of Status Relation
The fault map refers to a class of directed graph that takes the fault modes as the nodes and the relationship among failure modes as the directed edges, which can be expressed as $GN = \langle V_N, R_N \rangle$. In the formula, V_N is the collection of failure modes. The failure modes have different primary faults in the system as well as performance failure; it can be either real fault form of component or subsystem, or abnormal deviation of certain parameters in the system. R_N is the collection of binary relation among fault modes. It mainly describes the causal transmitting relations among various statuses. The status could be divided into system status and component status. System status represents the overall state of the system. It firstly represents the overall state of the system, that is, all the states of the system. The overall state could be decomposed into several substate, and the sub-state could be decomposed into sub-substate, thus forming a state layer structure based on state relations. In the layer structure, any lower layer is contained in state of the upper layer, and the upper layer state is constituted by its lower layer states. Collection can be used to represent this relationship: the bottom states serve as the basis to set up a cluster, and any system state is an element of the cluster, which is expressed as $Q(i)$, and the system state collection $\{Q\}$ is a subset of the cluster. If i is the upper state of j, then $Q(j) \subset Q(i)$. In order to meet the needs of automated analysis and matching of fault graph, this relation will be described as a tree structure.

Definition 4.3   State root tree

Directed root tree $T=<V, R>$ and its nodes correspond to the elements of system state collection $\{Q\}$, that is, node v_i corresponds to $Q(v_i)$ and it meets the requirements; if $r(v_j, v_i)R$, then $Q(v_j) \cap Q(v_i) = \varnothing$. And the directed tree T is the state root tree of state collection $\{Q\}$.

In other words, the state root tree describes the state structure in the form of a tree. The root node represents the complete set of system state, and the leaf node is the basic set of system state. There are three kinds of single phase power in the control system of the launch vehicles, and insulated grounding system is adopted between power supplies and between the power supplies and the launch vehicle's rocket shells. The electric power bus above the ground is $\pm M_1$, $\pm M_2$, and $\pm M_3$ respectively, and that on the rocket is $\pm B_1$, $\pm B_2$, and $\pm B_3$. Before the launch of the vehicle, examinations have to be carried out under the following 10 states respectively.

State 1: diversion insulation examination;
State 2: close sighting window;
State 3: subsystem;
State 4: control and telemetry match;
State 5: match of control, telemetry and cryogenic power system;
State 6: overall examination II-overall state examination;
State 7: overall examination I-overall state examination;
State 8: overall examination III-overall state examination;
State 9: general filling;
State 10: launch.

Meanwhile, component and unit that access the control system under different states varies with each other, and the state weight value of electric leakage of component under different conditions also varied. The troubleshooting of the test project of between two states should be evaluated based on the former state. For example, according to the above analysis, we can establish the power-supply model of negative bus of power supply, I (as shown in Figure 4.23).

Definition 4.4   State tree

The node set of directed tree T is V, and for the state set $\{Q\} = \{Q_B\} + \{Q_S\}$. If $V_1 \subset V$ and the following condition is met:

1. The node of V_1, together with the inner edge of V_1 in directed tree T, forms a layer root tree of $\{Q_S\}$, and the nodes with zero outdegree still has zero outdegree in tree T.
2. $V - V_1$ corresponds to $\{Q_B\}$ one by one and $\forall v - v_1, d^-(v) \neq 0$.
3. $\forall v \in V$, if the edge $r(v_i, v_j)$ exists, then $P(v_j) \subset P(v_i)$; if $P(v_j) \subset P(v_i)$, then there must be a path $v_j \rightarrow v_i$.

Briefly speaking, the state tree refers to a tree structure that describes the component states in a parallel way with the system state root tree at the core.

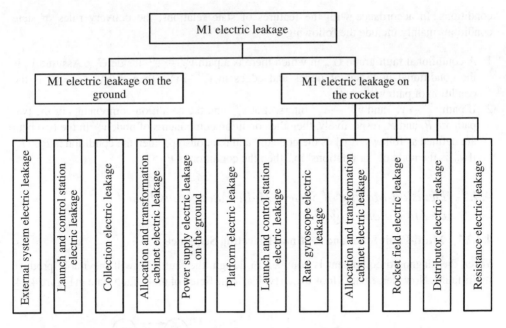

Figure 4.23   Hierarchical model of M1 electric leakage troubleshooting.

Structure of Conditional Fault Chart

> **Definition 4.5   Conditional fault chart**
>
> A directed graph $G_k = \langle V_k, R_k \rangle$ and a state tree $T_u = \langle V_u, R_u \rangle$, if V_K has two types of nodes, that is, $V_k = V_N + V_T$, and for any $v_i \in V_T$, the only correspondent node is $v_u \in V_U$ enabling v_t to satisfy the state relation between v_u. And it also meets the following requirements:
>
> 1. Node set V_N and its internal edge form a fault graph G_{KN}.
> 2. $\forall v_T^{(i)} \in V_T$, $v_N^{(i)} \in V_N$, there is no $\left(v_T^{(i)}, v_N^{(i)} \right) \in R$.
> 3. If $\left(v_T^{(i)}, v_N^{(i)} \right) \in R$, then there are two condition connection rules: if node $v_N^{(j)}$ has no access point in node set V_N, then node $v_T^{(i)}$ is the conditional node of node $v_N^{(j)}$, meaning that node $v_T^{(i)}$ is the existence condition of node $v_N^{(j)}$; if node $v_N^{(j)}$ has access point $v_N^{(g)}$ in node set V_N, then node $v_T^{(i)}$ is the conditional node of edge $v_N^{(g)}$, $v_N^{(j)}$, meaning that node $v_N^{(j)}$ is the condition for direct connection between $v_N^{(g)}$ and $v_N^{(j)}$, and the directed graph G_K is the conditional fault graph for the establishment of conditional tree T_u.

The integration of state condition node in the fault graph enables the fault communication to be condition communication. In the process of communication, the overlapping of conditions forms new condition communication and the overlapping or junction becomes the new conditions. In accordance with the features of state relations, the delivery rules of state conditions mainly include the following:

1. A conditional fault graph G_K, in which there is a path $v_N^{(i)} \to v_N^{(j)}$ inside G_{KN}. Assume V_T is the conditional node set of nodes and edges in $v_N^{(i)} \to v_N^{(j)}$, then $\cap P(V_t) | v_t \in V_T$ is the condition of path $v_N^{(i)} \to v_N^{(j)}$.
2. If path $v_N^{(i)} \to v_N^{(j)}$ and $v_N^{(g)} \to v_N^{(j)}$ intersect at $v_N^{(i)}$, and the condition components of the two path are P_1 and P_2 respectively, then the condition components of node $v_N^{(i)}$ in the two path are defined as follows: if $v_N^{(i)}$ is the node of "and relationship," then the condition is $P_1 \cap P_2$; if $v_N^{(i)}$ is the node of "or relationship," then the condition is $P_1 \cap P_2$.

4.4.2 Decision-Making of System Multi-Fault Status Based on Ant Colony Algorithm

4.4.2.1 Multi-Fault State and Optimal Detecting Sequence

When the diagnosed equipment is complex in design, the fault tree will inevitably produce multi-faults. Now please take a look at as a typical logic figure of multi-fault state in Figure 4.24.

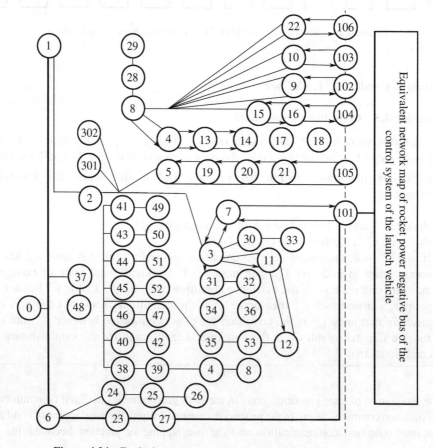

Figure 4.24 Equivalent network map of power supply I negative bus.

The tree has a number of n minimum fault sets. One fault symptom P corresponds to n fault detection paths $l_1, l_2, \ldots, l_n$, and there are a number of n possible fault reasons $R_1, R_2, \ldots, R_n$. The ith testing path has a number of m_i detection points $T_1, T_2, \ldots, T_n$. We can determine R_i, the ith fault reason, by testing l_i, the ith path. However, the process might give rise to two problems:

1. How to arrange the sequence of $l_1, l_2, \ldots, l_n$ of the n testing paths in fault diagnosis?
2. For a given path l_k, how to decide the sequence of its m_k testing points $T_{ki}, (j = 1, 2, \ldots, m_k)$?

Based on our experience, it is not suggested to determine the two kinds of sequences casually for the following reasons:

1. Different fault source (minimum cut set) has different importance and the fault source of greater importance should be given priority in fault diagnosis;
2. The difficulty of detection of different testing points also varies, and testing point, which is easier to detect, should be given priority.

Therefore, we must take into account the importance and difficulty of fault detection to acquire the optimal test sequence. For the m_k detecting points of a given path l_k, the detection should be carried out from easier detecting point to the harder ones, and if a fault condition is detected, you can confirm the fault source R_k and immediately suspend testing; otherwise, you can continue until the detection of m_k detecting points is completed. If the m_k detecting the point faults fails to meet the fault conditions, you can exclude the possibility of fault source R_k, and transfer to other testing paths for testing. This optimization, based on expert knowledge or practical experience, is easy to implement. The key of optimization is to decide the optimal sequence of $l_1, l_2, \ldots, l_n$ of the n testing paths. For this purpose, the ant colony algorithm is introduced in this section.

4.4.2.2 Dynamic Optimal Path Selection Based on Ant Colony Algorithm

The ant algorithm has global optimization, parallel distribution processing, and other characteristics. Here this section would like to emphasize the self-adaptability of the ant algorithm. When a barrier stands in the way of the ant's food-seeking process, the ants not only bypass the obstacles but also find a shortest path by observing changes of different paths via the ant pheromone trail after a period of positive feedbacks. With the ant colony algorithm, it is easy to find the optimization sequence of detection paths.

Ant Colony Algorithm Model
The ant-cycle model is a global optimized ant algorithm. If the pheromone trail density of i, j at moment t is $p(0 \le p \le 1)$, the updated formula of trail density is:

$$t_{ij}(t+1) = pt_{ij}(t) + \sum \Delta t_{ij}^k(t) \tag{4.35}$$

Assume Z_k is the length of the path of k ant in this cycle in which Q is a constant. If we assume η_{ij} is the length of path (i, j), the relative importance of path visibility is β $(\beta \ge 0)$, then the relative importance of the trail is α ($\alpha \geq 0$). U is the feasible vertex set, and P_{ij}^k, the transfer probability of ant k at moment t, is defined as:

$$P_{ij}^k(t) = \begin{cases} \dfrac{[t_{ij}(t)]^\alpha [\eta_{ij}]^\beta}{\sum\limits_{l \in U}[t_{il}(t)]^\alpha [\eta_{il}]^\beta}, & j \in U \\ 0, & \text{otherwise} \end{cases} \tag{4.36}$$

MMAS model (Max–Min Ant system) makes improvements on three points based on the ant algorithm (AS). First, we set the maximum value $t_{\max}$ of each pheromone trail in order to fully search for optimized sequence of each pheromone trail; second, we only modify the pheromone of the ant that goes on the shortest path in a circle, which is similar to the AS ant circle model adjustment; third, to ensure the algorithm finds the optimal solution, we constraint the pheromone concentration within $[t_{\min}, t_{\max}]$, and any value beyond this range will be set at $t_{\min}$ or $t_{\max}$. From the result of the experiment, MMAS algorithm has remarkably improved AS algorithm in terms of preventing algorithm prematurity and effectiveness.

Algorithm Design of Optimal Path Selection

Assume that the difficulty of T_{ij}, jth testing point on ith path l_i is $d(i, j)$, $(i = 1, \ldots, n, \ j = 1, \ldots, m_i)$. As we need to determine fault source R_i, we have carried out a maximum of m_i tests and a minimum of one test; therefore, the maximum cost of the path detection is:

$$\text{Max}(i) = \sum_{j=1}^{m_i} d(i, j) \tag{4.37}$$

the minimum cost:

$$\text{Min}(i) = \min\{d(i,1), d(i,2), \ldots, d(i, m_i)\} \tag{4.38}$$

The pheromone concentration released by unit ant is:

$$Q(i) = \frac{1}{\text{Max}(i) + \text{Min}(i)} \tag{4.39}$$

Obviously, the easier the path detection is, the more likely the ant would choose this path, and the pheromone released by the ant will increase correspondingly.

Assume the pheromone concentration of ith path at the moment of t is $\tau(t, i)$, and the importance of the path to the fault source R_i is $\eta(i)$. If we take both the pheromone concentration and the importance of fault source into account in the path selection, we could define the probability that the ant chooses ith path l_i at the moment of t as given in the formula (4.36).

Assume $G(V, E)$ is the weighted simple connected graph, in which $V = \{v_1, v_2, \ldots, v_n\}$ is the vertex set, $E = \{e_1, e_2, \ldots, e_n\}$ is the edge set, and n is the total number of vertices. $A = (a_{ij})_{n \times n}$ and $B = (b_{ij})_{n \times n}$ represent the connection matrix, distance matrix, and detection difficulty limit matrix respectively. Values of the connection matrix A are within 0–1 matrix. If the value of a_{ij} is 1, then there is edge between (i, j). If there is no edge between (i, j), the value of b_{ij} and c_{ij} is 0; if there is edge between (i, j), b_{ij} and c_{ij} are given a fixed value. m means the ant population and the objective function is as follows:

$$\min Z = \sum_{i=1}^{n} \sum_{j=1}^{n} b_{ij} a_{ij} \tag{4.40}$$

where s.t. $\begin{cases} c_{ij} a_{ij} > 0 \\ c_{ij} > 0 \\ a_{ij} > 0 \\ (i, j) \in E_x \\ E_x \subset E \end{cases}$.

4.5 Analysis of Intelligent Sneak Circuit

The launch vehicle composed of the components and parts → stand-alone → subsystem → system integration is a high technology involving complex structure and multiple fields of science and technology. In order to successfully test and launch the rocket in the launch site, the test and launch process must follow the stringent rules. One of the reasons for this is to avoid the impact of the launch vehicle on the sneak circuit. Studies on space mission failure in the past decade found that nearly 10% of failures are caused not by the device component failure nor the condition or man-made causes, but the defect in the product design, which is the sneak circuit. If sneak circuit failure occurs in rocket test and launch, the invisibility of the fault will bring great difficulties to fault analysis and removal. Such failures could affect the test plans, or even lead to equipment damage, launch failure, and loss of life.

At present, the SCA technology has two main categories: comprehensive SCA technologies and simplified SCA technologies. In 1967, the NASA of the United States proposed sneak analysis technique to improve the safety and reliability of the Apollo Moon program. This technique was put forward by Boeing as a standard analysis method, which attempted to reveal all functions and behaviors of the system. Therefore, the technique is also called "comprehensive SCA technologies." The disadvantage is that it has to be equipped with system-adaptive network tree generation algorithms and topology identification algorithm, and this technique can only be utilized after the completion of design and prior to production. Even if a sneak circuit is identified, it would bring enormous cost for modifications to the design of the entire system. It is these factors that limit the widespread of SCA technology.

The simplified SCA is divided into two types: SCA and design defect analysis. The SCA would search out all the paths between the "source" and "destination point" and apply two types of clues to identify the sneak circuit. The design defect analysis identifies design flaws through component level clues, which has no direct connection with design defect and source/destination points, serving as a supplement to the SCA. The two have been successfully applied in the relevant fields. However, SCA also has problems in the application. For example, the technique is error-prone when there are heavy workloads in the electronic systems input; great difficulties are faced in acquiring the threaded list; it is difficult to establish the model for complex components and produce false sneak circuit reports.

Targeting the complexity and heavy workloads of the traditional SCA, a neural network-based SCA utilizes the system-related information to form training samples, trains the neural network, and uses the trained neural network to predict all the functions of the system being turned on and off to identify potential problems by comparison with the design features.

This section first describes and analyzes all sneak circuits and specific issues in aerospace test and launch in detail, and summarizes general methods used in SCA of aerospace test and launch. Then the sections will discuss the vector machine-compatible SCA method based on BP network and its application in the SCA of aerospace test and launch.

4.5.1 Sneak Circuit Evaluation Rules and Analysis Techniques

When analyzing the sneak circuit malfunction, it is necessary to understand the evaluation rules at first. The simplest single network, as figure shown in 4.25, can be evaluated by the following three rules: (i) When load L is in need, whether switch S is in disconnect state; (ii) When load L is not in need, whether switch S is in connected state; (iii) Whether switch S could reflect real functions of load L, that is, when load L accesses or deviates from the circuit, whether switch S can indicate the state accurately. If any of the descriptions above is met, then there is a sneak circuit within the system.

The SCA technology would divide a huge and complex disordered systems figure and simplify it into a set composed of nodes and edges (branches); through ergodic path derivation and tracking, it produces network tree reflecting system connectivity, and by analyzing and evaluating the network tree, it decides whether there is a flaw with the sneak circuit.

The analysis of the sneak circuit is usually carried out through three steps: (i) Formation of a network tree, (ii) determination of the topology graph, and (iii) evaluation of the sneak circuit.

1. Formation of the network tree: This is the first step of SCA. For simple circuits, it is possible to form the tree with manual interference and analysis; for the composition for complex circuits, computer is needed to form the tree through automatic processing.
2. Determination of the topology graph: After the formation of the network tree, you can start identifying and classifying the basic topology graphics of each tree. The network of any circuit could be expressed by the combination of the following five basic topology graphics (see Figure 4.26). You can determine whether there is a sneak circuit according to the combination of the location of switch S and other identities in the network tree.
3. Evaluation of the sneak circuit: According to the topology graphics and the thread list associated with each graphic in combination with the sneak circuit rules, analyzing the circuit with the computer-aided system and combining with the observations of the researcher to finalize the state of the sneak circuit, all possible design flaws are so as to be discovered.

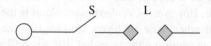

Figure 4.25   Single network diagram.

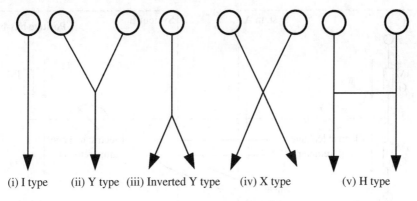

(i) I type   (ii) Y type   (iii) Inverted Y type   (iv) X type   (v) H type

Figure 4.26   Five basic topology graphics.

4.5.2   Potential Hazard and Analysis of Sneak Circuit in Aerospace Test and Launch

Sneak circuit may affect space test and launch in various degrees. Therefore, when there is the sneak circuit in the rocket launch site of the launch vehicle, a rapid analysis and processing as well as effective measures are required. So, in the failure analysis, we need to carry out rapid preliminary location and simplification of a complex circuit in order to improve the speed. In General, the sneak circuit can be divided into sneak circuits, sneak timing, sneak indicators, and sneak potential identifications. Now the section will analyze the hazards of a sneak circuit in space test and launch, mainly from the following four aspects.

4.5.2.1   Sneak Circuit

The sneak circuit refers to the phenomenon in which the current, energy, or logic flows along the unexpected channel or direction. The current launch vehicle has the most sneak circuits, mainly on the interface of the internal stand-alone of the system and interfaces between systems. This sneak circuit has a great effect on test and launch, and the analysis concerned is complicated—for simple circuits, artificial methods of analysis can be taken and for a complex circuit, the computer-aided system is usually required for analysis. When analyzing the sneak circuit, it is necessary to make clear the circuit of the system interface, the workflow, and operating principles to sort out such sneak circuits between the system interfaces.

Case study: During a mission, system A and system B of a certain type of launch vehicle failed in the electric leakage detection but qualified in electric leakage detections in turn. In the process of troubleshooting, the two systems were found unqualified in electric leakage detection only when tested simultaneously. Therefore, the detector suspected the problem was in the interface circuit. In analysis, the two systems went through electric leakage detection circuit and interface simplification (see Figure 4.27). It can be seen that the negative bus of the two systems was connected.

Because the electric leakage detection method for the two systems is in the same way, if the two systems are detected simultaneously, then it is possible that the negative bus of the other system would generate leakage current circuit (such as the sneak circuit in Figure 4.27), which would affect the two systems, leading to failed electric leakage detection.

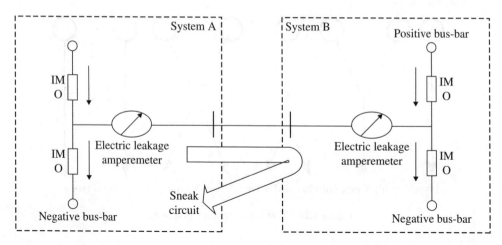

Figure 4.27   Interface between system A and system B on the ground.

4.5.2.2   Sneak Timing

Sneak timing refers to the phenomenon that events occur in unexpected or conflicting orders. When looking for this kind of problems, we mainly turn to data collection and data comparison to analyze the differences under the same operation state, a time difference of operation and operation timing differences, and then find out the problem.

Case study: When implementing matching examination for a task, system A was found to receive extra time strings and there is no anomaly record in the redundant monitoring or photographs and records by other systems when it lost a time string. By comparison, it was found that the system has changed its automated control mix line and analysis of the changed circuit was carried out. The cause of the failure was concluded as follows: due to route changes, the time string receiver threshold voltage did not match with the signal to be measured, resulting in sneak timing problems. The solution was as follows: adjust the ground mix threshold voltage so that it worked properly.

4.5.2.3   Sneak Indicators

The sneak indicator refers to the incorrect or implicit display of system working status. Guided by such incorrectness, operation staff may take unnecessary actions. This kind of problem is more obvious, which is mainly due to human error and can be avoided by following strict management regulations. As the launch vehicles have to undergo rigorous experiment and test in the factory, there is usually no flaw in sneak indicators. So far, there has been no evidence of this problem.

4.5.2.4   Sneak Identification

Sneak identification refers to incorrect or ambiguous identification of system condition. There are rarely identification errors in the launch vehicle. However, as different types of launch vehicles share the same set of ground equipment, the problem of vague identification still remains. The incorrect or ambiguous identification would lead to confusion under particular circumstances and cause sneak identification failures. There are two types of measures to prevent such failures: first, increase fault tolerance measures in function, and second, clear identification.

Case study: In the propellant loading preparation stage of a mission, system A is responsible for blow-down. As the ambiguous identification, the blow-off pipes of pump A and B on the ground are mistakenly connected to the blow-off pipe D1 and D2 of the rocket (the right connection should be pipe A with pipe D1 of the rocket); thus, the pipeline marked "Pump A blow-down" was connected to pipe D2 of the rocket while the pipeline marked "Pump B blow-down" was connected to pipe D1 of the rocket. This mistake led to a 2-hour delay of normal procedures, which had a great impact on the successful completion of the mission. There are two causes of the fault: one is that the specifications of D1 and D2 are completely the same and can even be exchanged from the physical perspective, thus causing confusion. Second is the identification inconsistency in which the blow-off pipes on the ground are marked as "Pump A blow-down" and "Pump B blow-down," whereas the identities on the rocket were "D1" and "D2". This inconsistency would easily lead to mistakes. Measures to be taken are to adjust the specifications of the two pipes to ensure that different pipes cannot be connected and identify the rocket-ground system in a consistent way.

4.5.2.5 Analysis of Typical Examples of an Aerospace Sneak Circuit

Analyze the secondary truncation circuit of system A of a certain type of launch vehicle with general analysis technology of a sneak circuit.

1. Data collection: As shown in Figure 4.28, when fully connecting the secondary truncating signal based on the actual demand, you need to disconnect the power transfer relays J1 and J2, and when the secondary truncating signal is disconnected, relay J1 and J2 remain in the

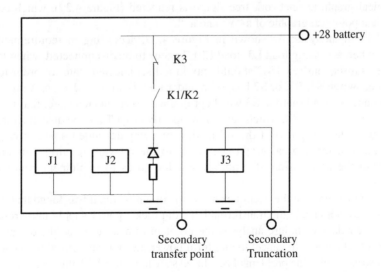

Figure 4.28 Circuit diagram of secondary truncation.

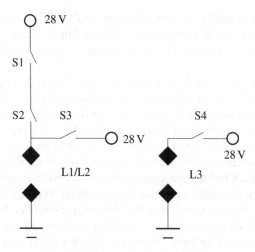

Figure 4.29   Network tree diagram.

disconnected state. In the overall test, the secondary turning signals are connected signals given by other systems, enabling actuation of power transfer relay KJ and K2. The correspondent K1 and K2 are switched from the disconnected to the connected state. Thereby it will mistakenly enable the power transfer function of the secondary device of the launch vehicle; the secondary truncating signal is a pulse-signal given to *m* from the system, actuating K3 relay, which disconnects the correspondent connected switch K3 so as to enable the truncation of a secondary device on the rocket. In the overall test, when the secondary flight is over, the system cuts off the power of the secondary devices of the rocket by sending secondary power-off signal.

2. Formation of a network tree: According to Figure 4.28 and by the method of drawing the topological graph, the network tree figure is rendered (Figure 4.29), which consists of the basic topological graphic of a "Y" and a "I."

3. Sneak circuit analysis: As shown in Figure 4.29, according to requirements of the design, when accessing load L3, load Ll/L2 needs to be disconnected; when load L3 is removed subsequently, L1/L2 should stay in a disconnected state. In order to achieve this state, switch S3, S1, or S2 has to be under the disconnected state. Analysis of real circumstances found that the S3 was kept in a wrong connected state, that is, S3 could not be kept disconnected, which caused incapacity of L1/L2 to be disconnected, inconsistent with the design. And the utilization of computer-aided analysis requires the rendering of a topological graph according to the relevant requirements and a clue sheet with the preparation of topology graphic so as to implement computer-aided analysis.

4. Processing measures: In the overall examination of a mission, it was found that the second truncation cannot enable the transformation from rocket power supply into ground power supply. After days of failure analysis, the problem turned out to be the design flaw, the solution of which was to cancel the uniform power transfer and turn to the system power transfer so as to solve the problem of S3 closed state in Figure 4.29, thus enabling secondary truncation.

The sneak circuit might seriously affect the test process and test quality of the launch vehicle and restrain improvement and reengineering of the rocket test process. Because of the complex system of the launch vehicle, analysis of the launch vehicle with sneak circuit technique cannot utilize artificial methods but computer-aided analysis methods to enable thread list automated analysis of the problem. In addition, to solve problems in design, the SCA is also needed in non-electrical system and test flight software. The SCA technology adopts the sequence of steps from inside the subsystems → between subsystems → the entire system to further improve the launch vehicle's design quality and eventually to the greater quality of rocket test and launch.

4.5.3   Neural Network-Based Analysis of Sneak Circuit

As a simpler and more effective method, the combination of neural network and the SCA is a new approach. The neural network-based analysis of a sneak circuit does not require detailed circuit and thread list but rather uses qualitative models and design capabilities of the system to analyze the system's sneak circuits.

4.5.3.1   Descriptions of Qualitative Models and Design Capabilities of the System

The component model consists of the terminal (component input/output), the internal structure (connection between component terminals, which may include a variable resistor), dependencies (changes of variable resistor under specific circumstances), external state (representing information between the components), and failure modes. The core part of the model is the variable resistance. The variable resistance defines three types of resistances, namely 0, l, and ∞ and $0 < l < \infty$. In order to apply the model to the neural network, now define qualitative resistance as 0, l, 1, $0 < l < 1$. 0 means short circuit, l means load, and 1 represents an open circuit. The circuit design function is $f_i \in F, 1 < i < m$, and m indicates the number of the functions, in which F is the design function space. f_i is valued 0 or 1. 0 indicates that the function is not implemented and 1 indicates that the function is realized. Assume that the system consists of n elements: $\{k_1, k_2, \ldots, k_n\}$, and the design function space is $\{f_1, f_2, \ldots, f_n\}$. For a given combination of components, the system should implement the appropriate combination of the functions.

4.5.3.2   BP Neural Network

The neural network structure is shown in Figure 4.30, which adopts the structure of three-layer BP neural network. The number of neurons in the input layer is the number of components of the system and in the output layer is the number of design functions, and the number of neurons at the intermediate layer can be determined by experience.

Hidden layer output:

$$Z_j = f\left(\text{net}_j\right), \quad j = 1, 2, \ldots, p$$

$$\text{net}_j = \sum_{i=1}^{M} \omega_{ij}^{(1)} x_i, \quad j = 1, 2, \ldots, p \tag{4.41}$$

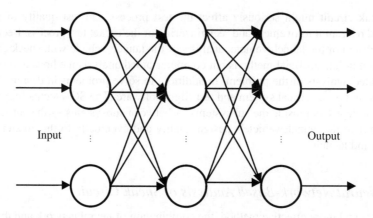

Figure 4.30   Structural diagram of neural network.

Output layer output:

$$y_k = f\left(\mathrm{net}_k\right), \quad k = 1, 2, \ldots, N+1$$

$$\mathrm{net}_k = \sum_{j=1}^{p} Z_j \omega_{jk}^{(2)} x_i, \quad j = 1, 2, \ldots, N \tag{4.42}$$

where $\omega_{ij}^{(1)}$ represents the link weight of input layer neuron i and the intermediate layer neuron j, and $\omega_{jk}^{(2)}$ means the link weight of intermediate layer neuron j and output layer neuron k, x_i stands for No. i neuron input, $f(\cdot)$ stands for activation function.

The learning algorithm is as follows:

$$\omega_{ij}^{(1)}\left(l+1\right) = \omega_{ij}^{(1)}\left(l\right) + \eta \Delta \omega_{ij}^{(1)}\left(l\right)$$

$$\omega_{jk}^{(2)}\left(l+1\right) = \omega_{jk}^{(2)}\left(l\right) + \eta \Delta \omega_{jk}^{(2)}\left(l\right)$$

$$\Delta \omega_{jk}^{(2)}\left(l\right) = \frac{\partial E}{\partial \omega_{jk}^{(2)}} = f'\left(\mathrm{net}_k\right) \sum_{j=1}^{p} f\left(\sum_{i=1}^{M} \omega_{jk}^{(2)} x_i\right) \tag{4.43}$$

$$\Delta \omega_{ij}^{(1)}\left(l\right) = \frac{\partial E}{\partial \omega_{ij}^{(1)}} = f'\left(\mathrm{net}_i\right) \sum_{j=1}^{p} \omega_{jk}^{(2)} f\left(\sum_{i=1}^{M} \omega_{ij}^{(1)} x_i\right) \sum_{i=1}^{M} x_i$$

4.5.3.3   SCA Theory Based on Neural Network

The main thoughts of neural network applied in SCA include the following: Firstly train the neural network through the relation between system design features and switch combination with the neural network input as the combination of system component state and functions as the output. Then utilize the trained neural network to predict functions enabled by all switch combinations. Identify the sneak circuits by comparing the differences between predicted functions and design functions.

Suppose the system component combination state is $\{k_1, k_2, \ldots, k_n\}$, in which n represents the number of system components, and the total number of combinations of the system is 3^n. The switch components are the main control elements in the circuit, of which there are only two states. Therefore, under normal circumstances, we care about a number of state combination 2^m, in which m means the number of system components. The design function combination is $\{f_1, f_2, \ldots, f_l\}$, in which l represents the number of design functions, f_l is represented in the form of probability, and $f_l = \{0, 1\}$. 0 indicates that the function is not implemented and 1 on the contrary. Train the neural networks through the combination of the design component states and the functional combination. After the training, the neural network can accurately predict the status of the design components and enabled functions. However, as to the untrained samples, the neural network may produce function $0 < f_l < 1$. In this case, results are predicted with threshold filtering. Thresholds β and predicting function have the following relationship:

$$f_l = \begin{cases} f_l = 1, f_l > \beta \\ f_l = 0, f_l \leq \beta \end{cases}, \quad 0 < \beta < 1 \tag{4.44}$$

After filtering, the SCA is implemented by comparison between the predicted results produced by neural network filtering and design functions. The value of β is closely related to the observed results of a sneak circuit and the changes of the value of β may lead to very different results. As the neural network output represents the function probability, $\beta \geq 0.5$ in most of the cases is required. According to the system design requirements, the neural network would predict functions enabled under all switch combinations, followed by comparison with the design functions to identify sneak circuits.

4.5.4 Sneak Circuit Analysis Based on Support Vector Machine

There are some specific problems in the application of method of sneak circuit based on the BP neural network, such as selection of threshold issue, the generalization ability of neural network, and problem explanation. The generalization ability of SVM outsmarts the traditional neural network method by improving the effectiveness of SCA to a large extent.

SVM originates from the development of statistical learning theory, a universal learning algorithm for small sample learning. There are mainly two types of SVM: classification and regression. Although the traditional SVM regression applies only to the case of single variable, the construction of a series of a single-variable support vector machine model could handle the complex multivariable system.

4.5.4.1 Multi-Variable SVM Regression Algorithm

Assume the collected samples include $(x^1, y^1), (x^2, y^2), \ldots, (x^r, y^r)$, $x^i \in R^d$, $y^i \in R^m$, where d and m are the number of independent variables and outputs respectively. Establish multivariate output model based on these samples using SVM regression. First, start with the simple linear model to explain some basic concepts of multivariate regression of SVMs.

Set f_i, $i = 1, 2, \ldots, m$ is the linear mapping, namely $f_i(x) = \langle \omega_i, x \rangle + b_i$, $\omega_i \in R^d$, $b_i \in R$, a $i = 1, 2, \ldots, m$ ($\langle , \rangle$ for vector inner product). According to Vapnik's structural risk minimization principle, $f_i(*)$ should enable the minimum

$$\frac{1}{2}\sum_{i=1}^{m}\|\omega_i\|^2 + C\sum_{i=1}^{m}\sum_{j=1}^{r}L_i\left(f_i\left(x^j\right), y_i^j\right) \tag{4.45}$$

where C is the equilibrium factor and $L_i(\cdot)$ is the loss function. The loss function $L_i(\cdot)$ usually uses ϵ-insensitive zone function, which is defined as:

$$L_i\left(f_i\left(x^j\right), y_i^j\right) = \begin{cases} 0, & \left|f_i\left(x^j\right) - y_i^j\right| < \epsilon_i, \epsilon_i > 0 \\ \left|f_i\left(x^j\right) - y_i^j\right| - \epsilon_i, & \text{others} \end{cases}$$

And the problem of multivariable SVM regression transformed into the optimization of the following problem:

$$\min J = \frac{1}{2}\sum_{i=1}^{m}\|\omega_i\|^2 + C\sum_{i=1}^{m}\sum_{j=1}^{r}L_i\left(f_i\left(x^j\right), y_i^j\right)$$

$$\text{s.t. } \left|f_i\left(x^j\right) - y_i^j\right| < \epsilon_i \tag{4.46}$$

However, under these constraints, because the optimization problem does not necessarily have a solution, the slack variables are introduced in order to guarantee beingness of solutions. And the optimization problem (4.46) can be written as:

$$\min J = \frac{1}{2}\sum_{i=1}^{m}\|\omega_i\|^2 + C\sum_{i=1}^{m}\sum_{j=1}^{r}\left(\xi_i^j + \xi_i^{j*}\right)$$

$$\text{s.t.} \begin{cases} y_i^j - \langle \omega_i, x^i \rangle - b_i \le \epsilon_i + \xi_i^j \\ \langle \omega_i, x^i \rangle + b_i - y_i^j \le \epsilon_i + \xi_i^{j*} \\ \xi_i^j, \xi_i^{j*} \ge 0 \end{cases} \tag{4.47}$$

Introduce the Lagrange function

$$L = \frac{1}{2}\sum_{i=1}^{m}\|\omega_i\|^2 + C\sum_{i=1}^{m}\sum_{j=1}^{r}\left(\xi_i^j + \xi_i^{j*}\right)$$

$$- \sum_{i=1}^{m}\sum_{j=1}^{r}\alpha_i^j\left(\epsilon_i + \xi_i^j - y_i^{j*} + \langle \omega_i, x^j \rangle + b_i\right)$$

$$- \sum_{i=1}^{m}\sum_{j=1}^{r}\alpha_i^{j*}\left(\epsilon_i + \xi_i^{j*} + y_i^j - \langle \omega_i, x^j \rangle - b_i\right)$$

$$- \sum_{i=1}^{m}\sum_{j=1}^{r}\gamma_i^{j*}\left(\xi_i^{j*} + \xi_i^j\right)$$

$$\tag{4.48}$$

The extreme value L of the Lagrange function should meet the following conditions:

$$\frac{\partial L}{\partial \omega_i} = 0, \qquad \frac{\partial L}{\partial b_i} = 0,$$

$$\frac{\partial L}{\partial \xi_i^j} = 0, \qquad \frac{\partial L}{\partial \xi_i^{j*}} = 0,$$

$$i = 1, 2, \ldots, m; \quad l = 1, 2, \ldots, r.$$

And it is concluded that

$$\begin{cases} \dfrac{\partial L}{\partial \omega_i} = \omega_i - \displaystyle\sum_{j=1}^{r} \alpha_i^j x^j + \sum_{j=1}^{r} \alpha_i^{j*} x^j = 0 \Rightarrow \\[2mm] \omega_i = \displaystyle\sum_{j=1}^{r} \left(\alpha_i^j - \alpha_i^{j*} \right) x^j \\[2mm] \dfrac{\partial L}{\partial b_i} = -\displaystyle\sum_{j=1}^{r} \alpha_i^j + \sum_{j=1}^{r} \alpha_i^{j*} = 0 \Rightarrow \\[2mm] \displaystyle\sum_{j=1}^{r} \left(\alpha_i^j - \alpha_i^{j*} \right) = 0 \\[2mm] \dfrac{\partial L}{\partial \xi_i^j} = C - \alpha_i^j - \gamma_i^j = 0 \\[2mm] \dfrac{\partial L}{\partial \xi_i^{j*}} = C - \alpha_i^{j*} - \gamma_i^j = 0 \\[2mm] i = 1, 2, \ldots, m; \quad l = 1, 2, \ldots, r \end{cases} \qquad (4.49)$$

Introduce formula (4.49) into the Lagrange function (4.48), and we get the dual form of the optimization problem

$$\max W \left(\alpha_i^j, \alpha_i^{j*} \right) = -\frac{1}{2} \sum_{i=1}^{m} \sum_{l,j=1}^{r} \left(\alpha_i^j - \alpha_i^{j*} \right) \left(\alpha_i^j - \alpha_i^{l*} \right) \left\langle x^j, x^l \right\rangle$$

$$+ \sum_{i=1}^{m} \sum_{l,j=1}^{r} \left(\alpha_i^j - \alpha_i^{j*} \right) y_i^j - \sum_{i=1}^{m} \sum_{l,j=1}^{r} \left(\alpha_i^j + \alpha_i^{j*} \right) \epsilon_i \qquad (4.50)$$

$$\text{s.t.} \sum_{l,j=1}^{r} \left(\alpha_i^j - \alpha_i^{j*} \right) = 0$$

$$0 < \alpha_i^j, \alpha_i^{j*} < C$$

computing α_i^j and α_i^{j*}, and the expression of regression function $f_i(x)$, $i = 1, 2, \ldots, m$ is obtained as follows.

$$f_i(x) = \left\langle \omega_i, x \right\rangle + b_i$$

where $\omega_i = \displaystyle\sum_{j=1}^{r} \left(\alpha_i^j - \alpha_i^{j*} \right) x^j$.

According to Karush–Kuhn–Tucker (KKT) conditions, the optimal solution meets the following conditions:

$$\begin{cases} \alpha_i^j \left[\epsilon_i + \xi_i^j - y_i^{j*} + f_i\left(x_j\right)\right] = 0. \\ \alpha_i^{j*} \left[\epsilon_i + \xi_i^{j*} + y_i^j - f_i\left(x_j\right)\right] = 0. \\ i = 1,2,\ldots,m; \quad l = 1,2,\ldots,r. \end{cases} \tag{4.51}$$

$$\begin{cases} \gamma_i^j \xi_i^j = 0, \\ \gamma_i^{j*} \xi_i^{j*} = 0, \\ i = 1,2,\ldots,m; \quad l = 1,2,\ldots,r. \end{cases} \tag{4.52}$$

From formula (4.51), it can be seen that $\alpha_i^j \alpha_i^{j*} = 0$, that is, both α_i^j and α_i^{j*} are zero. From formula (4.52) and (4.49), we get

$$\begin{cases} \left(C - \alpha_i^j\right) \xi_i^j = 0, \\ \left(C - \alpha_i^{j*}\right) \xi_i^{j*} = 0, \\ i = 1,2,\ldots,m; \quad l = 1,2,\ldots,r. \end{cases} \tag{4.53}$$

By analysis of formula (4.53), it is known that only when $\alpha_i^j = C$ and $\alpha_i^{j*} = C$, the error of $f_i(x)$ and y_i might be above ϵ_i; therefore,

$$\begin{cases} \epsilon_i - y_i^{j*} + f_i\left(x_j\right) = 0, \quad 0 < \alpha_i^j < C \\ \epsilon_i + \xi_i^{j*} - f_i\left(x_j\right) = 0, \quad 0 < \alpha_i^{j*} < C \\ i = 1,2,\ldots,m; \quad l = 1,2,\ldots,r. \end{cases} \tag{4.54}$$

According to formula (4.54), we could compute the threshold variable.

$$\begin{cases} b_i = y_i^j - \left\langle \omega_i, x^j \right\rangle - \epsilon_i = 0, \quad \alpha_i^j \in (0,C) \\ b_i = y_i^j - \left\langle \omega_i, x^j \right\rangle + \epsilon_i = 0, \quad \alpha_i^{j*} \in (0,C) \\ i = 1,2,\ldots,m; \quad l = 1,2,\ldots,r. \end{cases} \tag{4.55}$$

It is possible to work out the threshold variable with either of the SVMs, and averaging approach also applies.

4.5.4.2   Nonlinear Regression Algorithm of Multivariate SVM

For the nonlinear regression problem, we can use a nonlinear mapping $\Phi(\cdot)$ to map the trained data to a high-dimensional feature space and establish a linear regression function in this high-dimensional feature space. As the process of optimizing only considers high-dimensional inner product operation, the kernel function $K(x, y)$ alone on behalf of the inner product operation $\langle \Phi(x), \Phi(y) \rangle$ in the high-dimensional space could produce nonlinear regression. Just like the derivation of linear regression above, we obtain a formula similar to (4.50) with the same constraint conditions, which is shown as below.

$$\max W\left(\alpha_i^j, \alpha_i^{j*}\right) = -\frac{1}{2}\sum_{i=1}^{m}\sum_{l,j=1}^{r}\left(\alpha_i^j - \alpha_i^{j*}\right)\left(\alpha_i^l - \alpha_i^{l*}\right)K\left(x^j, x^l\right)$$
$$+ \sum_{i=1}^{m}\sum_{l,j=1}^{r}\left(\alpha_i^j - \alpha_i^{j*}\right)y_i^j - \sum_{i=1}^{m}\sum_{l,j=1}^{r}\left(\alpha_i^j + \alpha_i^{j*}\right)\epsilon_i \tag{4.56}$$

Its constraints conditions are the same as the optimization formula (4.51). And the kernel function is used to calculate the inner product of characteristics spaces:

$$K\left(x^j, x^l\right) = \langle \Phi(x), \Phi(y) \rangle = \Phi(x)\Phi(y) \tag{4.57}$$

The multivariate nonlinear regression function could be expressed as:

$$f_i(x) = \sum_{j=1}^{r}\left(\alpha_i^j - \alpha_i^{j*}\right)K\left(x_i, x\right) + b_i, \quad i = 1, 2, \ldots, m. \tag{4.58}$$

Like the problem of linear regression, α_i^j and α_i^{j*} are worked out by optimization formula (4.56), and the threshold variable b_i is worked out by KKT conditions.

$$\begin{cases} b_i = y_i^j - \sum_{l=1}^{r}\left(\alpha_i^l - \alpha_i^*\right)K\left(x_i, x\right) - \epsilon_i = 0, & \alpha_i^j \in (0, C) \\ b_i = y_i^j - \sum_{l=1}^{r}\left(\alpha_i^l - \alpha_i^{l*}\right)K\left(x_i, x\right) + \epsilon_i = 0, & \alpha_i^{j*} \in (0, C) \\ \quad i = 1, 2, \ldots, m; \qquad\qquad\qquad j = 1, 2, \ldots, r. \end{cases} \tag{4.59}$$

4.5.4.3 Analysis of Practical Examples

Figure 4.31 is the firing circuit of the Redstone rocket of the United States. As a typical case of a sneak circuit, it can be seen from Figure 4.31 that after the ignition switch is closed, and if the tail grounding plug is pulled out prior to tail pull-off plug, the problem of a sneak circuit will occur as shown in dotted lines in the figure, that is, the ignited missile would cause the undesired shutdown.

There are four controllable components in the system: ignition switch, emergency shutdown switch, tail pull-off plug, and tail grounding plug. The expected functions of the system include ignition and shutdown, and the corresponding functional components are the shutdown coil and

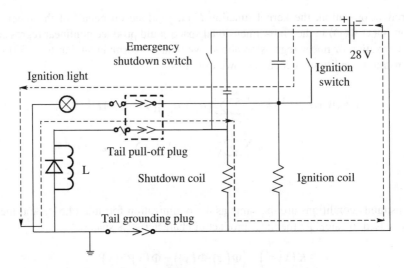

Figure 4.31   Simplified circuit of redstone rocket firing.

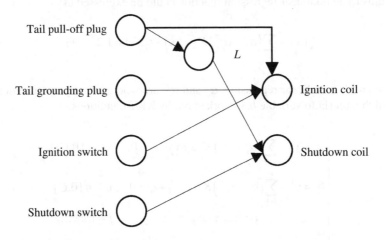

Figure 4.32   Firing network model of redstone rocket.

the ignition coil. In the circuits, power source and the ignition light are normal components; inductance coil L and the capacitor are intermediate components. The emergency shutdown switch and ignition switch are directly connected to the power source while the shutdown coil and ignition coil are directly connected to the ground. The converted network model is shown in Figure 4.32.

Design requirements: when the ignition switch is turned off, the ignition coil is enabled and the tail grounding plug and the tail pull-off plugs are opened at the same time; when turning off the shutdown switch, the shutdown coil works. According to the design requirements, a system design matrix is created and shown in Table 4.5. In the table, 1 indicates that the switch is closed or the component is enabled and 0 indicates that the switch is opened or the component does not work.

Table 4.5 Firing circuit design matrix of the Redstone rocket.

No.	Ignition switch	Shutdown switch	Tail pull-out plug	Tail grounding plug	Ignition coil	Shutdown coil
1	1	0	1	1	1	0
2	0	1	0	0	0	1
3	0	0	0	0	0	0

Table 4.6 Training sample.

No.	Input 1	Input 2	Input 3	Input 4	Output 1	Output 2
1	0	0	0	0	0	0
2	0	0	0	1	0	0
3	0	0	1	0	0	0
4	0	0	1	1	0	0
5	0	1	0	0	0	1
6	0	1	0	1	0	1
7	0	1	1	0	0	1
8	0	1	1	1	0	1
9	1	0	0	0	1	1
10	1	0	1	1	1	0
11	0	0	1	0	0	0
12	1	1	0	0	1	1
13	1	1	0	1	1	1
14	1	1	1	0	1	1
15	1	1	1	1	1	1

Because of the single state of the functional components in the circuit, (0, 1) is used for the purpose of description. From the circuit, we can see that when the shutdown switch is closed, the shutdown coil is enabled; when the ignition switch is closed, the ignition coil works, and the tail pull-out plug and tail grounding plug are disconnected simultaneously. However, in practice, the manufacturing errors may cause asynchronously disconnection of the tail pull-out plug and tail grounding plug. The history data shows that when the tail pull-out plug and tail grounding plug are disconnected at the same time, the system continues to function properly. In addition, although it is not allowed to close the shutdown switch and ignition switch at the same time, we can form a virtual sample of this state for training SVMs. Through the analysis above, a training sample can be obtained as shown in Table 4.6.

The experiment first targets three different BP networks whose structure is 4-10-2, 4-5-2, and 4-15-2, respectively. In order to improve the generalization ability of the network, the experiment adds white noise into the samples to form 150 new samples and trains the network with 120 samples, while the remaining 30 samples are taken as test samples. The generalization errors of the three BP networks are 0.22, 0.41, and 0.64 respectively. As the switch circuit has only two states, 0 and 1 are utilized to indicate them. The threshold β is set as 0.5. Because the four states has a total of 16 different combinations, Table 4.6 presents 15 combinations, while the predicting function of the left switch combination $\{1, 0, 0, 0\}$ is unknown. The three trained BP neural networks predict the functions of the state combination as $\{1, 0.80\}$, $\{0.51, 0.45\}$, and $\{0.42, 0.43\}$. So the implemented functions are $\{1, 1\}$, $\{1, 0\}$, and $\{0, 0\}$. Through analysis, it can be determined that the output of the latter two BP networks is incorrect. So it is crucial to guarantee the generalization capability of BP networks so as to ensure the effectiveness of the output.

Then train the SVM with samples. In the selection of kernel functions, the radial basis function as below is adopted:

$$k\left(x, x_i\right) = \exp\left(-\frac{\|x_i - x_i\|_2^2}{\sigma^2}\right)$$

where $\sigma = 10$. The parameters of the SVM are $\epsilon = 0.01$, $C = 500$.

It can be found from the experiment that the SVM can accurately determine the functions of switch combination $\{1, 0, 0, 0\}$, and its generalization capability is higher than the single BP network, thus improving the reliability of the analysis.

4.6 Malfunction Diagnosis Based on Recurrent Neural Network

The dataset of aerospace products collected in the process of emission test is often a small sample one. The electromagnetic interference would impede the completeness of the data acquired sometimes, in addition to other interferences induced by a specific environment, which usually occurs only once or is unlikely to appear again, making it even harder to detect malfunctions during the emission test. In the dynamic recurrent neural network, Hopfield neural network is quite effective in handling small samples, incomplete and noise interfered-data due to its unique associative memory. For this reason, a malfunction diagnosis approach based on Hopfield neural network will be discussed in this section. It is noted that this section targets analog circuit on spacecraft, which could also be applied to other targets.

4.6.1 Associative Memory and Hopfield Neural Network

In his classic papers, Hopfield proposes a theory that a physical system composed of a number of simple elements (neurons) features emergence of collective behaviors. In short, the collectivity of a system is not demonstrated by a single element, but rather the interaction between local units in a system.

4.6.1.1 Associative Memory

Memory stands as an instinctive feature of human beings, which plays a part in most cognitive models. For this, any memory system involving association is referred to as associative memory. Associative memory requires a system to output a complete expectation associated mode when inputting a mode into the system. The associative memory of neural network, unlike the traditional computers that store every input/output mode, needs only to store the shifting mechanism of input/output. In order to explain associative memory, a few basic concepts are to be illustrated at first.

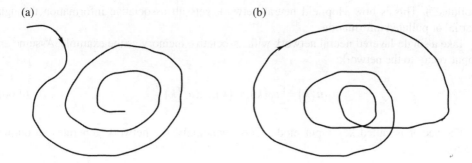

Figure 4.33   Network attractors. (a) Fixed point attractor and (b) limit cycle attractor.

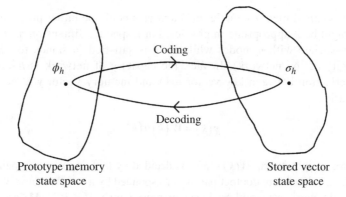

Figure 4.34   Coding/decoding by Hopfield neural network.

Attractor: A state when the network is stable. In the state space, the attractor may be a single point which is called a fixed (stable) point; it may also refer to a periodic evolution between different states under some principle, which is called limit cycle, as shown in Figure 4.33.

Domain of Attraction: A collection of initial states that stabilizes the network under the same attractor is called the domain of attraction of the attractor.

Prototype state (prototype memory) ϕ_h is represented by fixed (stable) point σ_h of the dynamic system. Therefore, ϕ_h could be reflected on stable point σ_h of the network. The reflection represents $\phi_{h \leftarrow} \sigma_h$, of which the forward direction (from left to right) stands for coding and the backward direction (from right to left) decoding. Figure 4.34 explains the coding/decoding of state space of Hopfield neural network associative memory. In the recalling process, a key mode is handed to the network, an input corresponding to the initial point of the state space. Assume that the input mode contains all the information (or partial information) of a table point of the prototype memory state space, then the dynamic system would evolve with time into a memory state when the initial point is "close" to the stable point of the memory searches, thus stimulating the system to generate proper memory.

There are a few fixed attractors (or stable equilibrium point) that does not evolve with time in the Hopfield neural network. When establishing a memory mode as a stable equilibrium point of the network, the system could associate with the mode when the stable equilibrium point starts from an initial state nearby (which could also be interpreted as "the mode is polluted"). This is how Hopfield neural network gets all associative information through partial or polluted information.

Take a single-layered neural network with associative memory as the example. Assume an input vector to the network:

$$u(k) = \left[u_1(k), u_2(k), \ldots, u_n(k) \right]^T \qquad (4.60)$$

The vector is called key input mode. Correspondingly, the network generates an output vector:

$$y(k) = \left[y_1(k), y_2(k), \ldots, y_n(k) \right]^T \qquad (4.61)$$

The vector is called memory mode. $u(k)$ and $y(k)$ could be either positive or negative, although it might be inappropriate in physics. For a specific dimension n, the neural network could associate with h modes while $h \leq n$ is satisfied (n stands for the maximum storing capacity of the network). In fact, the capacity of network is $h < n$. The linear associative reflection between key vector $u(k)$ and memory vector $y(k)$ could be represented by matrix:

$$y(k) = W(k)u(k) \qquad (4.62)$$

In the matrix, weight matrix $W(k) \in R^{n \times n}$ is decided by both input/output pair $\{u(k), y(k)\}$, where $k = 1, 2, \ldots, h$. Each input/output pair is corresponded by a weight matrix $W(1)$, $W(2)$, ..., $W(h)$. The weight matrix set could produce a memory matrix $M \in R^{n \times n}$. M describes the total of weight matrix of each input/output pair (or the whole associative mode set). Therefore, memory matrix M could be represented by:

$$M = \sum_{k=1}^{h} W(k) \qquad (4.63)$$

The memory matrix could define the overall connection between any input mode (key mode) and related output (memory mode). Plus, the memory matrix could be regarded as a representation of collective experience acquired by the network through h input/output modes. The recursive form of Formula (4.63) is represented as:

$$M_{k+1} = M_k + W(k), \quad k = 1, 2, \ldots, h \qquad (4.64)$$

In the matrix M, $M_0 = 0$. The results from a Formula (4.63) match with that of Formula (4.64). When the memory matrix is built from matrix incremental $W(k)$ produced by k-numbered input/output pairs $\{u(k), y(k)\}$, the current weight matrix $W(k)$ would lose its individuality by mixing with the previous $(k-1)$ weight matrix. However, all the information would not be lost in the mixing process for the current associative message, which could be estimated through specific key mode and related memory mode by the memory matrix.

4.6.1.2 Discrete Hopfield Neural Network

The time-discrete Hopfield neuron network adopts McCulloch–Pitts model to build its neural structure and hard limiter function for its activation function. Therefore, the state of the network at any time could only be −1 or +1. The topological structure of the network is shown in Figure 4.35.

For each neuron in Figure 4.35, the input of linear combiner is represented as:

$$v_j(k) = \sum_{i=1}^{n} w_{ji} x_i(k) - \theta_j(k) \quad j = 1, 2, \ldots, n \tag{4.65}$$

In the formula, $x(k) = [x_1(k), x_2(k), \ldots, x_n(k)]^T$, representing the state of network; $\theta_j(k)$ stands for j-numbered external threshold of neural. For $j = 1, 2, \ldots, n$, the output of each linear combiner is conveyed to symmetrical hard limiter activation function and elements of unit delay. The unit delay output $x_j(k)$ is the input from other neurons apart from its own. By setting $i=j$, the process gets $w_{ji} = 0$. Therefore, the state of each neuron could be represented by:

$$x_j(k+1) = \text{sgn}(v_j(k)) = \begin{cases} +1 & v_j(k) > 0 \\ -1 & v_j(k) < 0 \end{cases} \tag{4.66}$$

If $v_j(k) = 0$, $x_j(k)$ is defined as the value of the previous moment. If the external threshold is represented as vector form, that is, $\theta(k) = [\theta_1(k), \theta_2(k), \ldots, \theta_n(k)]^T$, the matrix output of network is:

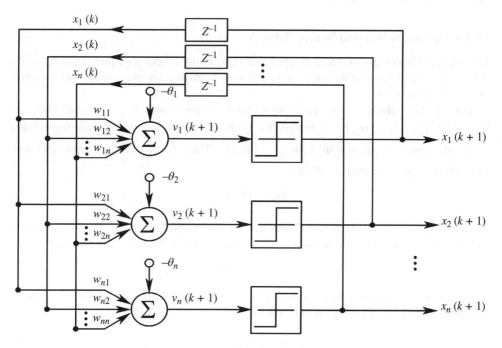

Figure 4.35 Discrete Hopfield neural network model.

$$x(k+1) = \mathrm{sgn}\left(Wx(k) - \theta(k)\right) \tag{4.67}$$

In the formula, W stands for memory matrix M, as shown in formula (4.64). According to customs of neural network descriptions, it is described as W:

$$W = \begin{bmatrix} 0 & w_{12} & \cdots & w_{1n-1} & w_{1n} \\ w_{21} & 0 & \cdots & w_{2n-1} & w_{2n} \\ \vdots & \vdots & \ddots & \vdots & \vdots \\ w_{n-11} & w_{n-12} & \cdots & 0 & w_{n-1n} \\ w_{n1} & w_{n2} & \cdots & w_{nn-1} & 0 \end{bmatrix} \tag{4.68}$$

The associative memory of time-discrete Hopfield neural network could be divided into two stages, storing stage and recalling stage. In the storing stage, we assume the network has a total of r memory prototypes, that is, $\{\phi_1, \phi_2, \ldots, \phi_r\}$. The network weight is calculated by the following formula:

$$W = \frac{1}{n}\sum_{i=1}^{r}\phi_h\phi_h^T - \frac{r}{n}I \tag{4.69}$$

In the recalling process, the tested input vector $x' \in R^{n\times 1}$ is conveyed to the network. Initialize the network state with unknown input $x(k)$, that is, $x(k)\big|_{k=0} = x(0) = x'$. Update network state asynchronously until it becomes stable. When this requirement is met, the stabilized (balanced) state x_e represents network output.

4.6.1.3 Quantum Hopfield Neural Network

This section will discuss quantum Hopfield neural network model with time delay based on the mechanism of associative memory in collation with lineal superposition and quantum measurement theories.

Assume that there is a key input mode represented by quantum state as $|\psi_k\rangle = \left[|\psi_{k1}\rangle, |\psi_{k2}\rangle, \ldots, |\psi_{kn}\rangle\right]^T$. Correspondingly, there is an output memory mode represented by quantum state as $|\Phi_k\rangle = \left[|\Phi_{k1}\rangle, |\Phi_{k2}\rangle, \ldots, |\Phi_{kn}\rangle\right]^T$. Then, for a specific input/output pair $\{|\psi_k\rangle, |\Phi_k\rangle\}$, they are related as follows:

$$|\Phi_k\rangle = W_k |\psi_k\rangle \tag{4.70}$$

In the formula, W_k represents the evolution matrix of quantum state. Then the memory matrix of quantum state could be built by the evolution matrix of quantum state, that is.

$$J_k = J_{k-1} + W_k \tag{4.71}$$

According to this theory, we get the quantum Hopfield neural network based on the similar Hopfield network by integral path of Feynman:

$$\Psi_m^{out} = \sum_{n=1}^{N} J_{mn} \Psi_n^{in} \tag{4.72}$$

In the formula, $J_{mn} = \sum_{k=1}^{P_s} \varphi_m^k \left(\varphi_m^k\right)^*$. P_s stands for the total number of network memory modes, $\left(\varphi_m^k\right)^*$ is the complex conjugate of φ_m^k, and m and n represent the line and row of evolution matrix J_{mn} respectively.

In the quantum Hopfield neural network model above, a few considerations are needed:

1. As quantum state evolution goes along with the time, it may be affected by time delay. In addition, feedback to the vectors of network input during the operation of Hopfield neural network is the result of time delay. Therefore, it is necessary to consider time delay in building up quantum Hopfield neural network model.
2. The result of the quantum state is captured with certain probabilities when measured. In the operation of quantum Hopfield neural network, an evolution matrix corresponding to a quantum state also appears on certain probabilities. In other words, when the quantum Hopfield neural network is built, the evolution matrix should appear in the form of a measurement matrix considering the effect of probability.

For this reason, this section proposes a quantum Hopfield neural network model with time delay, the network topology of which is shown in Figure 4.36. $\hat{O}$ is a superposition operator, representing the superposition of quantum states.

The quantum Hopfield neural network model in Figure 4.36 could be described as:

$$
\begin{aligned}
\left|x(t)\right\rangle &= \sum_{k=0}^{d}\sum_{j=1}^{n}\left\langle w_{ij}^k \middle| x_j(t-k)\right\rangle \\
&= \sum_{k=0}^{d} W^k \left| x(t-k)\right\rangle
\end{aligned}
\tag{4.73}
$$

In the formula, d is the number of steps of time delay. To make this model universally applicable, the input from any neuron should be composed of input with k step of time delay beforehand $k = 1, 2, \ldots, d$ in the quantum Hopfield neural network model with time delay. $W^k = \left(\left|w_{ij}^k\right\rangle\right)_{n \times n}$ represents the quantum state measurement matrix at k-numbered step of time delay. The reason of the elements of a quantum measurement matrix existing in the form of quantum state is that a quantum state is, in fact, a complex number (two-dimensional vector), and the weight connecting each quantum state should correspondingly be a complex number in order to enable interactions between each ground state of the quantum state.

The connecting weight matrix plays the most important part in building Hopfield neural network model. The composition of a measurement matrix is also a vital stage in the quantum Hopfield neural network model. According to associative memory mechanism, measurement matrix W^k and memory matrix W should satisfy the following:

$$
W = \sum_{k=0}^{d} W^k
\tag{4.74}
$$

It is noted that memory matrix W is by all the memory modes (assume the total number is h). Generally, the total number of memory modes h is not specifically related to time delay. However, the quantum measurement matrix under each delay state in the quantum

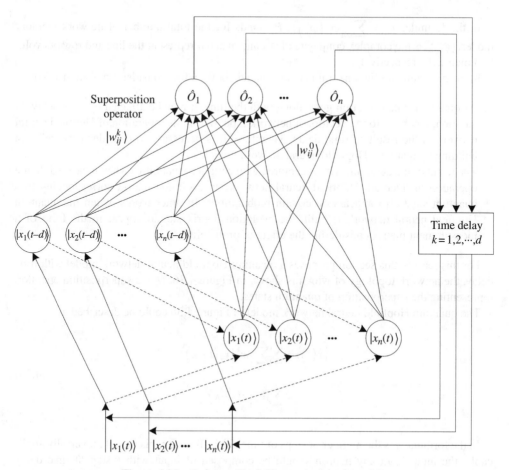

Figure 4.36 Quantum Hopfield neural network model.

Hopfield neural network also reflects the evolution mode of quantum evolution. To simplify this process, in this section two assumptions are made about the quantum Hopfield network model proposed: (i) the total step of time delay represents the total number of memory modes of the network, namely $h=d$ (so the summation factor in a formula (4.74) is d) and (ii) assume $d=1$, namely studying the model under just one step of time delay. Then when we assume the total number of memory prototypes in the network is r {$|x_1\rangle, |x_2\rangle, ..., |x_r\rangle$}, the relation between quantum measurement matrix and memory could be calculated according to the equation below:

$$W = \frac{1}{n}\sum_{i=1}^{r}|x_i\rangle\langle x_i| - \frac{r}{n}I \tag{4.75}$$

$$W = W^0 + W^1 \tag{4.76}$$

$$w_{ij}^0 = \begin{cases} \rho_i w_{ij} + \rho_j w_{ji} & i \neq j \\ 0 & i = j \end{cases} \tag{4.77}$$

$$\rho_i = \frac{\langle x_i(t) | x_i(t) \rangle}{\displaystyle\sum_{i=1}^{n} \langle x_i(t) | x_i(t) \rangle} \qquad (4.78)$$

where ρ_i means the quantum measurement factor.

As a matter of fact, apart from the quantum measurement matrix, we are more interested in the probability of getting correspondent quantum measurement matrix. We define this probability as P, which satisfies:

$$P^k = \sum_{i=1}^{n} \sum_{j=1}^{n} \left| p_{ij}^k \right| \qquad (4.79)$$

$$p_{ij}^k = \frac{w_{ij}^k - w_{ij}^{k-1}}{w_{ij}^{k-1}} \qquad (4.80)$$

4.6.2 Diagnosis Based on Hopfield Neural Network

4.6.2.1 Diagnosis of Hopfield Neural Network

The analog circuit diagnosis system based on Hopfield network works is shown in Figure 4.37. The system could be divided into two parts, namely data preprocessing and malfunction classification.

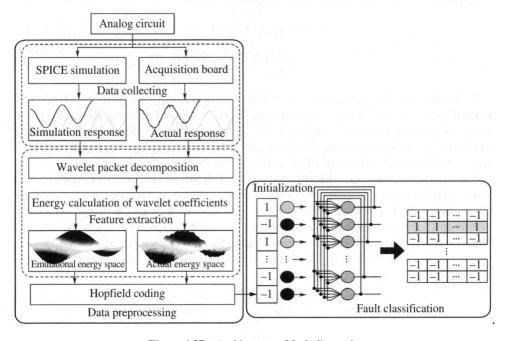

Figure 4.37 Architecture of fault diagnosis.

The data preprocessing module involves three sub-modules, namely data collecting, feature extraction, and Hopfield coding. In the data collecting module, the output response of a simulative circuit collects data samples through MULTISIM simulation and the data collecting board connected to the actual circuit terminal so as to get ideal output response data set and the actual output response data set. In the feature extraction module, the ideal and actual circuit output responses go through wavelet packet decomposition as training and testing data set respectively, and the feature vector of corresponding malfunction is made up of energy value through energy calculation based on the decomposed wavelet coefficients. In the Hopfield coding sub-modules, the feature vector of each sample is coded under specific rules. In the malfunction classification module, the malfunctions coded by Hopfield are handed to the Hopfield network so as to achieve accurate and rapid malfunction classification. Now we will introduce some key technologies involved.

Wavelet Packet Decomposition and Energy Calculation

The wavelet packet analysis is an extension of wavelet analysis mentioned in previous chapters. The wavelet analysis divides a circuit output response into lowpass filter (LF) and highpass filter (HF) respectively according to approximation coefficient and detail coefficient. The process is completed by scale transformation and translation transformation of mother wavelet, that is,

$$\psi_{a,b}(x) = \frac{1}{\sqrt{a}}\psi\left(\frac{x-b}{a}\right) \tag{4.81}$$

In the formula, $\psi_{a,b}(x)$ represents mother wavelet function, in which a and b donate the scale and translation factor of wavelet transform respectively. The wavelet coefficient of circuit response $I(x)$ could be represented by the following formula.

$$
\begin{aligned}
C(a,b) &= \left\langle \psi_{a,b}(x), I(x) \right\rangle \\
&= \frac{1}{\sqrt{a}} \int I(x)\psi\left(\frac{x-b}{a}\right)dx
\end{aligned}
\tag{4.82}
$$

To facilitate the calculation on the computer, discrete dyadic wavelet transforms is frequently adopted by the engineer. Adequate results could be obtained just by commanding scale factor and translation factor as $a = 2^j$ and $b=k$, $2^j = ka$, $(j, k) \in Z^2$. Inspired by Multi-Resolution Analysis, the wavelet packet analysis decomposed the circuit response $I(x)$ into the approximate part of LF and detail part of HF on each scale, as shown in Figure 4.38.

A number of statistics which describe the features of signals will be acquired through wavelet packet analysis of signals, such as mean squared error, histogram, covariance matrix, and energy value. Among the features above, the energy value acquired through calculations of HF/LF wavelet coefficient provides effective descriptions of features. Through this approach, we could divide the signal components of various frequencies in the energy space with a more detailed way.

In his technical report about feature extraction of time sequence by DWT and DFT, Morchen defines the energy value as:

$$
\begin{aligned}
E_{m.n}(x) &= \sum_{i=1}^{N}\left|C_i^{m,n}(x)\right|^2 \\
&= \left\|C_i^{m,n}(x)\right\|_2^2
\end{aligned}
\tag{4.83}
$$

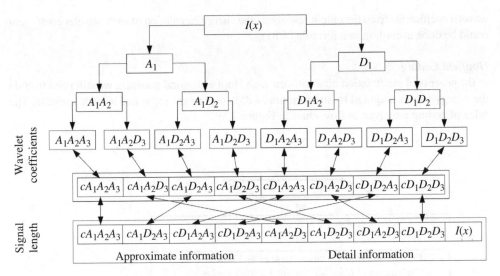

Figure 4.38   Wavelet packet analysis.

where m stands for the depth of wavelet decomposition, n stands for the number of nodes wavelet trees under the depth of m, and N is the length of the wavelet coefficients of a node (m, n). In fact, most of the current methods could calculate energy value based on this definition.

Ekici and other researchers promoted this definition on the basis above, that is,

$$E_{m.n}(x) = \sum_{i=1}^{N} \left| C_i^{m,n}(x) \right|^p$$

$$= \left\| C_i^{m,n}(x) \right\|_p^p$$

(4.84)

where $1 \leq p \leq \infty$.

In fact, the efficiency of energy definition that formula (4.83) describes in distinguishing signals when extracting their features is not satisfactory, especially for weak signals. For example, we get a very small (or very big) absolute value of energy value through the formula (4.83). Under the same wavelet nodes, each energy value is quite close to each other (the difference value between energy values is not sufficient enough for the purpose of valid distinguishing). In the energy calculation of formula (4.84), the researchers could get ideal p values only through their adequate experiences. And when the research target is changed, the selected p value before might not be suitable for feature extraction of current signals. For this reason, a new energy function for signal extractions will be defined in this section:

$$E_{m.n}(x) = \left\| C^{m,n}(x) \right\|_2^2 N^{-1} \sum_{i=1}^{N} \exp \left(-\frac{C_i^{m,n}(x)^2}{2} \right)$$

(4.85)

In the process of feature extraction, the wavelet decomposition is divided into three layers while Haar wavelet is adopted as the wavelet basis (db1 wavelet). After acquiring HF and LF wavelet coefficients from the output responses, the energy calculation of each wavelet coefficient could be done according to a formula (4.85).

Hopfield Coding

In the process of malfunction diagnosis through Hopfield neural network, we still need to code the energy values acquired by the formula (4.85) so as to get rapid and accurate results. The rules of coding are given as flow chart in Figure 4.39.

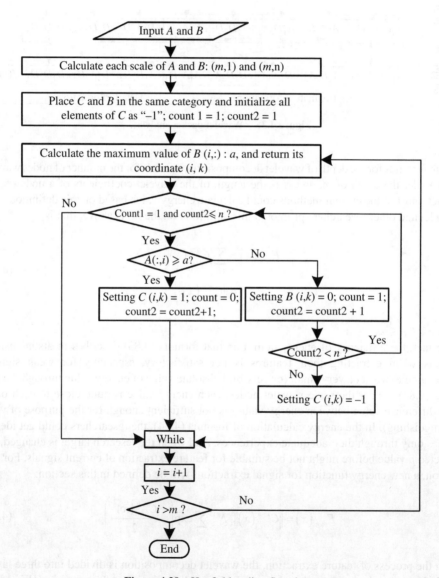

Figure 4.39 Hopfield coding flowchart.

In Figure 4.39, $A \in R^{m \times 1}$ represents the energy value vector of some actual malfunction response, and $B \in R^{m \times n}$ stands for a matrix composed of energy values of all ideal malfunction responses (m is the dimension composed by HF/LF wavelet coefficient energy values, and n represents the categories of malfunctions). $C \in R^{m \times n}$ is the corresponding Hopfield coding matrix of actual malfunction. When coding the energy value of a specific signal, the coding process would search for all energy values of malfunction responses that are placed in the same wavelet node as the signal in the energy space. Therefore, the coding rules could accurately reflect the coding state of energy values.

4.6.2.2 Malfunction Diagnosis of Quantum Hopfield Neural Network

In the diagnosis approaches under Hopfield neural network above, there are several issues that deserve specific considerations:

1. The malfunction diagnosis approach through Hopfield neural network resembles somewhat "searching in a dictionary." In other words, when a specific malfunction occurs, the malfunction mode is handed to the Hopfield network and "match" with the corresponding memory prototype driven by associative memory (the initial point of feature space converges to the stable point), through which the category of malfunction is decided. However, when the memory prototype lacks a corresponding mode to the malfunction, the diagnosis approach might fail. In this case, it is necessary to add corresponding memory prototype, resulting in the poor "flexibility" of the method.
2. In a real situation, the worker usually judges and evaluates the malfunction information acquired, going through a decision process with certain probabilities. Though the result of malfunction diagnosis is a determined value (signal feature values that exceed control limits or a determined number of a category) under approaches like multivariate statistics, neural network, and SVM, it does not necessarily guarantee the occurrence of the diagnosis result. Under certain situations, the determined results are usually accompanied by "false warning." Therefore, it is proper to understand the results as the probability of malfunction rather than its occurrence.

For this purpose, this section proposes a malfunction diagnosis approach of quantum Hopfield neural network with probability explanation mechanism. It is noted that the extraction of the malfunction energy features in this approach resembles those based on Hopfield neural network except for the following three aspects:

1. The diagnosis approach of quantum Hopfield neural network is based on the quantum evolution principles, which considers single malfunction as quantum ground state, multi-malfunctions as quantum excited state, and coupled multi-malfunctions could be represented as the probability superposition of single malfunction so as to explain the causes of multi-malfunctions from the probability perspective.
2. The malfunction diagnosis approach based on quantum Hopfield neural network requires no Hopfield coding of the acquired energy features. Instead, it quantizes the energy features.
3. The neural network in quantum Hopfield neural network diagnosis adopts quantum Hopfield neural network instead of the traditional Hopfield neural network.

The process of quantum Hopfield neural network diagnosis includes the following:

1. Quantize the ideal single malfunction energy feature and the actual energy features of multi malfunctions and go through orthonormalization.
2. Consider the preprocessed standard single malfunction energy features as quantum memory prototype and calculate the correspondent memory matrix according to formula (4.75).
3. Calculate the measurement matrix of each memory prototype according to formula (4.77) and formula (4.78).
4. Set the energy features of preprocessed actual multi-malfunctions as the key input mode of quantum associative memory and calculate the measurement matrix of each key input mode according to formula (4.77) and formula (4.78).
5. Calculate the correspondent measurement matrix of memory prototypes under each key input mode.
6. Calculate the probability matrix of the occurrence of memory prototypes under each key input mode according to formula (4.80).
7. Calculate the occurrence probability of each memory prototype under the key input modes based on the probability matrix in step 6.
8. Decide the probability of superposition of single malfunction in forming multi-malfunctions according to the probability value acquired through the step 7.

4.6.3   Case Study of Malfunction Diagnosis

4.6.3.1   Single Malfunction Diagnosis of an Analog Circuit

The faulty circuit mentioned in this section is a Sallen–Key band-pass filter as shown in Figure 4.40. The central frequency of the circuit is 25 kHz when all components work within the range of normal value. It is assumed that the resistance and capacitance have a tolerance of 5% and 10% respectively in the filter. The Sallen–Key filter would output a malfunction response when any part in R2, R3, C1, and C2 is 50% higher or lower than the normal value and the other parts in the circuit alter within the tolerance range. In this way, we could get eight

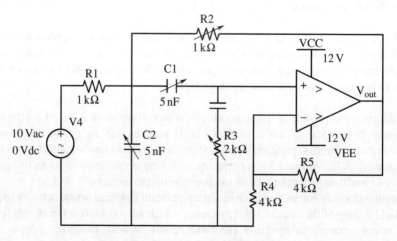

Figure 4.40   Sallen–Key band-pass filter.

types of malfunctions, namely R2⇑, R3⇑, C1⇑, C2⇑, R2⇓, R3⇓, C1⇓, and C2⇓, of which ⇑ and ⇓ represent 50% higher than the normal value and 50% lower than the normal value.

The Sallen–Key filter is posed upon an excitation response from an impulse sequence of 5 V amplitude, 10 μs pulse width. The ideal output response and the actual output response of the circuit under different malfunction modes are achieved through SPICE simulation and data collection from the data collection board in the terminal of the circuit respectively. The data collector is PCI-6289 with an analog input of 32 bits, resolution of 18 bits, and sampling frequency of 500 kS/second (multichannel). The period of the collection is 10 ms. Figure 4.41 describes the ideal and actual circuit data of Sallen–Key filter under different malfunction modes.

From Figure 4.41c, it is clear that there is not much difference between signal trends of normal output response and faulty output response. Therefore, the signal variance alone could not decide the occurrence of malfunctions. We need to preprocess the signals with certain analysis tools and extract relevant malfunction features so as to diagnose the malfunctions effectively.After getting circuit responses, we acquire necessary malfunction energy features through wavelet decomposition and calculations based on formula (4.85). The approximate wavelet coefficients and the energy values of the detail wavelet coefficients of Sallen–Key filter under three scales are shown in Table 4.7. In the table, the first subscript of E represents the decomposition scale and the second subscript is the number of wavelet nodes.

Each energy value alone offers limited information of malfunction and fails to serve as features in distinguishing different malfunction mode. Inspired by "collective emergence," the researchers combine the energy values of the relevant wavelet coefficients under each scale into one vector. It displays the nature difference between malfunctions under different modes through the interaction between each element of the vector. And we call the space composed of feature vectors the space of malfunction features. In this space, the standard malfunction feature vector (memory prototype) collected from ideal data set is unique while the individual malfunction feature vector (initial memory point) acquired through actual measurement would evolve into the memory prototype with the help of associative memory. The malfunction feature subspace of Sallen–Key filter is shown in Figure 4.42. In the space, F1, …, F8 represents C1⇓, C2⇓, R2⇓, R3⇓, C1⇑, C2⇑, R2⇑, and R3⇑ respectively and NF means no malfunction.

After calculating the malfunction energy feature subspace, we get the codes of Sallen–Key bass-band filter under different malfunction modes according to coding rules in Figure 4.39, as shown in Figure 4.43. In Figure 4.43, (3,0)–(3,7) represent eight types of malfunctions under coding state of corresponding wavelet node, namely C1⇓, C2⇓, R2⇓, R3⇓, C1⇑, C2⇑, R2⇑, and R3⇑.

In Figure 4.43, the malfunction features coding under real situation is considered as the "initial memory point" (network input) and is handed to the Hopfield neural network while the malfunction features coding of Sallen–Key filter under SPICE simulation (ideal situation) is regarded as "stable memory state" (network target vector) and is handed to the Hopfield network. After setting the proper time step (the time step proposed in this section is five steps), the actual malfunction feature, driven by the associative memory of Hopfield network, evolves toward the predetermined stable memory state (malfunction memory prototype). Thus rapid and accurate malfunction categorization would be completed. Figure 4.44 describes the diagnosis results of Sallen–Key filter with soft malfunctions through the approach proposed in this section. In Figure 4.44, F1, …, F8 represents C1⇓, C2⇓, R2⇓, R3⇓, C1⇑, C2⇑, R2⇑, and R3⇑, and (3,0)–(3,7) stand for the state of the corresponding wavelet nodes.

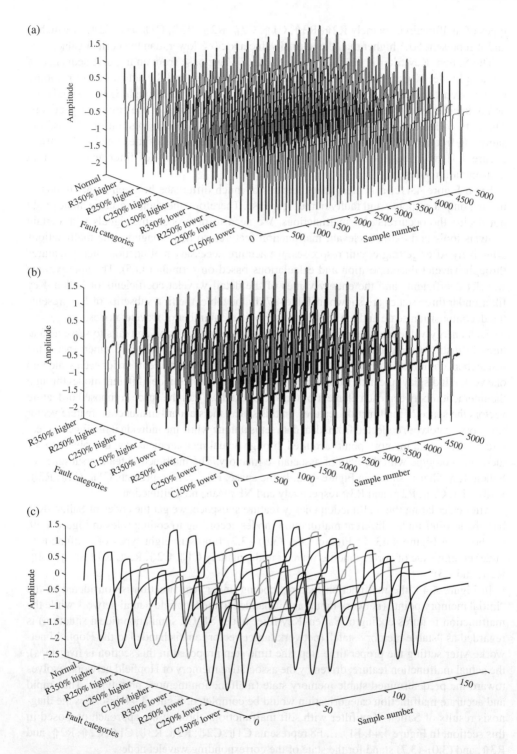

Figure 4.41 Output response of Sallen–Key bass-band filter. (a) SPICE simulated output response, (b) actual output response, and (c) partial enlarged view of diagram (a).

Table 4.7　Energy values of Sallen–Key bass-band filter.

Category of fault	$E_{3,0}$	$E_{3,1}$	$E_{3,2}$	$E_{3,3}$	$E_{3,4}$	$E_{3,5}$	$E_{3,6}$	$E_{3,7}$
Ideal C1⇓	29.6775	6.7007	3.3069	1.5585	1.6904	0.7651	0.4295	0.5090
Ideal C2⇓	33.5858	7.1649	3.5352	1.8253	1.7930	0.8604	0.4535	0.5452
Ideal R2⇓	36.2734	7.0677	3.5495	1.3374	1.7789	0.6458	0.3358	0.4294
Ideal R3⇓	26.1819	6.0872	2.9817	1.4669	1.5008	0.7203	0.4121	0.5073
Ideal C1⇑	29.3510	6.1503	3.0444	1.2170	1.5336	0.5935	0.3237	0.3948
Ideal C2⇑	28.4127	5.8698	2.9036	1.1380	1.4661	0.5631	0.2874	0.3512
Ideal R2⇑	29.8841	6.3127	3.1270	1.3044	1.5476	0.6166	0.2949	0.3861
Ideal R3⇑	32.7519	6.4958	3.2338	1.2398	1.6323	0.6093	0.3107	0.4070
Normal	31.7179	6.4925	3.2256	1.3083	1.6149	0.6253	0.3069	0.4019
Actual C1⇓	29.7287	6.9382	3.5391	1.6745	1.7850	0.6187	0.4458	0.5369
Actual C2⇓	35.3749	6.3082	3.5357	1.3215	2.0384	1.2781	0.5193	0.5078
Actual R2⇓	33.2558	7.1137	3.5400	1.3972	1.7847	0.6104	0.4051	0.4835
Actual R3⇓	27.5793	9.6395	3.0168	1.4085	1.5307	0.7255	0.3286	0.5082
Actual C1⇑	29.4064	6.2147	3.0682	1.3045	1.5348	0.7116	0.3301	0.3693
Actual C2⇑	28.5044	6.5728	2.9037	1.1429	1.5340	0.5689	0.3043	0.3601
Actual R2⇑	27.7924	6.3519	3.5497	1.3243	1.6200	0.6098	0.3055	0.3904
Actual R3⇑	33.0582	7.0688	3.2745	1.2177	1.6359	0.6094	0.4125	0.4162

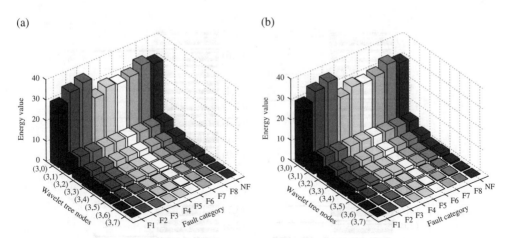

(a)　　　　　　　　　　　　　　　　　　(b)

Figure 4.42　Malfunction feature subspace of Sallen–Key bass-band filter. (a) Standard malfunction feature sub-space and (b) actual malfunction feature sub-space.

4.6.3.2　Multi-Malfunction Diagnosis of an Analog Circuit

This section continues to take Sallen–Key filter as an example to discuss malfunction diagnosis of coupling malfunctions based on quantum Hopfield network. Now consider the four situations of coupling malfunctions below: C1 and C2 are simultaneously 50% lower than the tolerance, that is, C1⇓ and C2⇓ (MF1); R2 and R3 are simultaneously 50% higher than the tolerance, that is, R2⇑ and R3⇑ (MF2); R2 and R3 are simultaneously 50% higher while that

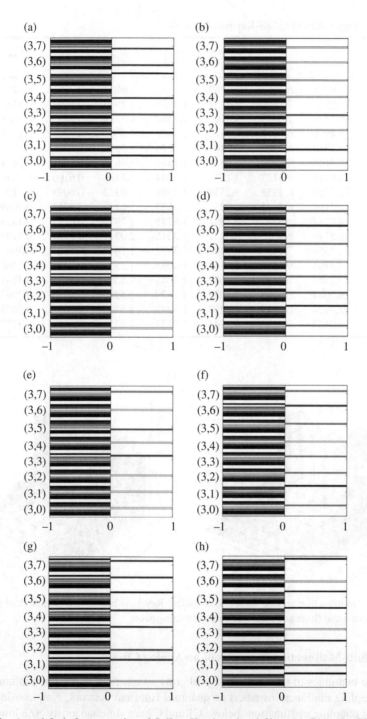

Figure 4.43 Actual fault feature codes of Sallen–Key bass-band filter. (a) C1⇓, (b) C2⇓, (c) R2⇓, (d) R3⇓, (e) C1⇑, (f) C2⇑, (g) R2⇑, and (h) R3⇑.

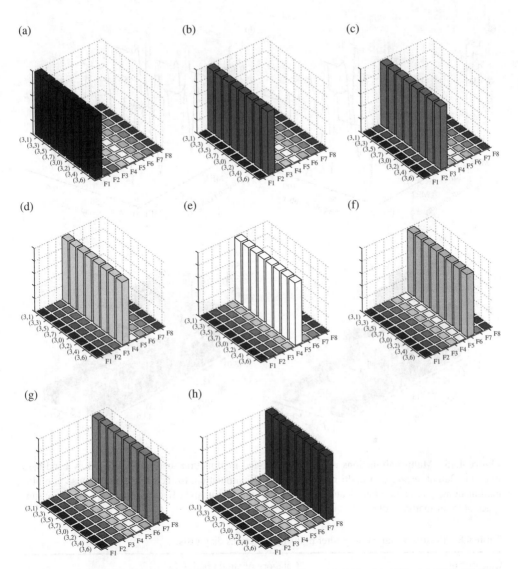

Figure 4.44 Fault categories of Sallen–Key band-pass filter. (a) C1⇩, (b) C2⇩, (c) R2⇩, (d) R3⇩, (e) C1⇑, (f) C2⇑, (g) R2⇑, and (h) R3⇑.

of C1 is 50% lower than the tolerance, that is, R2⇑R3⇑C1⇩ (MF3); R2 and R3 are simultaneously 50% higher while that of C1 and C2 are 50% lower than the tolerance, that is, R2⇑R3⇑C1⇩C2⇩ (MF4). Figure 4.45a and b describe the output responses of the four malfunctions under SPICE simulation and real situation (In the numerical experiment, we collected 5000 samples of responses from each coupling malfunctions. And the section offers the partially enlarging view of malfunction signals.) Through wavelet decomposition and energy calculation as mentioned in Section 4.6.2, we get the standard and actual energy space of four malfunctions, as shown in Figure 4.45c and d.

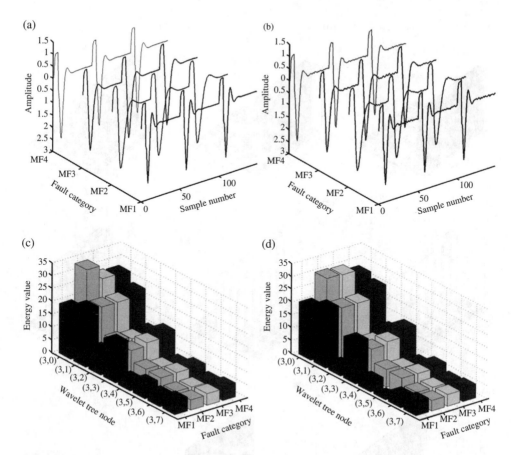

Figure 4.45 Multi-malfunctions output response and malfunction feature of Sallen–Key band-bass filter. (a) Output response of multi-malfunction SPICE simulation, (b) actual output response of multi-malfunctions, (c) standard fault feature sub-space of multi-fault, and (d) actual malfunction feature sub-space of multi-malfunctions.

Table 4.8 Results of Sallen–Key filter multi-malfunctions diagnosis.

Category of multi-faults	Category of single fault (%)							
	C1⇓	C2⇓	R2⇓	R3⇓	C1⇑	C2⇑	R2⇑	R3⇑
MF1	76.56	68.36	5.86	10.16	16.41	14.84	18.36	22.66
MF2	13.67	5.08	13.28	18.75	20.70	4.30	66.02	73.05
MF3	17.19	75.39	10.55	8.59	10.55	24.61	67.97	62.50
MF4	63.67	61.72	14.84	6.25	11.33	16.02	69.92	75.39

The occurrence probability of multi-malfunctions against single malfunction is shown in Table 4.8. It is clear that the occurrence probability of multi-malfunctions MF1 against single malfunction C1⇓ is 76.56%, and that of MF1 against single malfunction C2⇓ is 68.36%, higher than another single malfunction. Therefore, we could consider that multi-malfunction

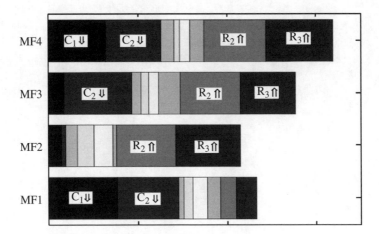

Figure 4.46   Probability distribution of Sallen–Key bass-band filter multi-malfunction diagnosis.

MF1 is the superposition of single malfunction C1⇩ and C2⇩. The result matches with the mode of multi-malfunctions coupling in real situation. Similarly, multi-malfunctions MF2 is the superposition of single malfunction R2⇧ and R3⇧ with a probability of 66.02 and 73.05%; multi-malfunctions MF3 is the superposition of single malfunction C2⇩, R2⇧, and R3⇧; and multi-malfunctions MF4 is the superposition of single malfunction C1⇩, C2⇩, R2⇧, and R3⇧ with a probability of 63.67, 61.72, 69.92, and 62.50%.

Figure 4.46 offers the distribution of probabilities of multi-malfunctions diagnosis results, in which the eight patches of different colors in MF1–MF4 stand for eight types of single malfunctions and the length of each patch means the probability of each single malfunction.

5

Safety Control of Space Launch and Flight—Modeling and Intelligence Decision

Safety control plays a vital part in space launch and flight technologies. The safety control of space launch and flight aims at monitoring effectively the safety channels, subsatellite point track, and flight parameters of the launch vehicle in the initial stage of space launch, the most important and the dangerous stage in the whole process. It also guarantees the safety of the vehicle and carries out safety control decisions in a timely fashion. Therefore, we need to build up a model to calculate parameters concerning safety control and relevant parameters during the flight, determine the flight state of the launch vehicle, and make accurate and reliable safety control decisions when the flight parameters reach or exceed the warning limits.

This chapter will introduce the calculations of safety control parameters such as real-time trajectory, safety channels, and impact point as well as safety deciding approaches and safety control strategies; discuss average approaches for flight path calculation through multisourced data fusion; illustrate how liquid propellant propels launch vehicle explosion under malfunction state, present the shell splintering model, toxic gas leakage, and diffusion model and Geographic Information System (GIS) calculation and analysis model; introduce the calculation of explosion range; and discuss the intelligent safety control decisions based on space information processing and loss estimation as well as intelligent emergency response decisions in face of malfunctions.

5.1 Overview of Space Launch and Flight Safety Control

Because of the complexity and risks of space launch technologies, the launch vehicle, once struck with malfunction in the process from launch to stage separation, would endanger the vital facilities to the impact point and surrounding cities with hundreds of tons of fuels it carries, bringing serious safety threats. It will inevitably exert huge losses on people's life, properties, and important facilities on the ground if handled improperly. Therefore, safety control, as a major measure for safety guarantee of space launch safety, is an essential aspect in a space launch.

Intelligent Testing, Control and Decision-Making for Space Launch, First Edition. Yi Chai and Shangfu Li.
© 2015 National Defense Industry Press. Published 2015 by John Wiley & Sons Singapore Pte Ltd.

In the process of space launch, we need to calculate parameters like the safety channels, sub-satellite point, and trajectory in real-time situations, decide the current stage of the launch vehicle in an accurate and reliable manner, and take necessary measures in the safety control system when real-time data reach or exceed warning limits. As a major measure to guarantee launch safety and fix malfunctions, safety control requires correct judgment and the ability to reflect the current flight state. Therefore, it is necessary to obtain, handle, and analyze the current trajectory and stage of a launch vehicle in an accurate and reliable way, calculate precisely parameters of flight trajectory and safety channel, guarantee the safety of launch flight, and carry out real-time decisions of safety control.

5.1.1   Basic Concept of Space Launch and Flight Safety Control

In the process of space launch and flight, safety control includes the generation and transmission of safety control decisions and safety controls orders. It deals with the real-time trajectory data of various measurement equipments in a real-time manner, compares the results of calculation with theoretical data and the safety channel, and estimates the state of the launch vehicle. When the real-time parameters reach or exceed the malfunction limits, a sign of malfunction of the launch vehicle, it will make safety control decision and implement the decision on the launch vehicle, terminating the powered flight and destroying the vehicle.

Safety control bears much significance for space launch and flight and serves as the most effective means to minimize losses in case of malfunction during the launch of the vehicle. It is clear to see that the reliability and accuracy of safety control decision lay as the basis for the ground safety control system to complete a launch mission successfully.

The safety control decision of space launch and flight has three features as below:

1. Comprehensiveness: The safety control takes not only each parameter concerning the flight of a launch vehicle into account but also geographical and environmental information, especially information about the impact points to estimate the current state of the launch vehicle. The information tends to be complex in category and enormous in amount.
2. Reliability: As safety control plays a vital part in ensuring people's life and property, it has to function properly during the whole process of space launch and flight but also acquire, handle, and analyze relevant information correctly, guaranteeing accurate results and precise reflection of the current state of safety control (such as the current state of the vehicle, ground circumstance).
3. Real-time capability: As the vehicle flies at high speed in the process of launch and flight, the safety control instructor may have little time to handle any malfunction. Therefore, the safety control system has to finish the process of information acquisition, processing, and analysis in a very short time to make the right decision.

5.1.2   Information-Processing Flowchart of Space Launch and Flight Ground Safety Control System

The safety control system of the launch vehicle includes the vehicle safety control system and the ground safety control system. The section will focus on the features of ground safety control system and proper approaches targeting the model of space launch and flight safety control and relevant intelligent decisions. The ground safety control system of launch vehicles includes the

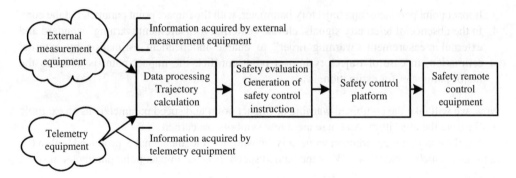

Figure 5.1   Information processing flowchart of safety control system.

whole process from the collection of vehicle information and estimation of flight state to the generation and delivery of safety controls orders, among which the real-time estimation of flight state stays at the core of the whole system. How to get accurate results and avoid faulty safety control decisions or any mistake is the key of flight safety control model and intelligent decisions. The information-processing flowchart of the system is shown in Figure 5.1.

The ground safety control system of a launch vehicle is mainly used to handle and monitor various flight parameters, estimate the flight state of launch vehicle, and send correspondent safety control orders. From the safety control system information processing flowchart in Figure 5.1, it is clear that the system takes external and telemetry equipment as the data basis. Proper data processing and accurate trajectory calculation ensure the reliability of further judgment while appropriate safety estimation rules and safety control model guaranteed the right results and safety controlled orders.

5.1.3   Approaches for Safety Estimation and Safety Control Strategies

Approaches for safety estimation and safety control strategies serve as references for the safety management and control of launch vehicles and flight. In the process of flight, the system decides whether the vehicle is in normal condition, mainly by real-time telemetry data that reflects the function of the vehicle as well as its flight trajectory. Parameters reflecting the functions of vehicle and its trajectory are used for safety control of the vehicle. The safety channel decides whether the vehicle has exceeded the limits, the warning line forms the boundary of the vehicle's powered flight, and the blowing line is the boundary that terminates the powered flight of the malfunction-stricken vehicle. When the ground safety control system implements destruction of the launch vehicle, it has to choose the proper impact point (impact point selection).

5.1.3.1   Approaches for Safety Estimation

When designing approaches for safety estimation, the principles below should be considered:

1. External and remote measurement, with external measurement at the core and telemetry as supplement.
2. Trajectory impact point parameter and telemetry parameter, with the former as the major player.

3. Impact point parameter and trajectory parameter, with the impact point parameter at the core.
4. In the absence of telemetry signals, elevate "external measurement warning" standard and external measurement "warning order" to enable independent estimation by external estimation; in case of trajectory-impact point conflict, the impact point is taken as the major parameter for estimation.

Now, we will take the combined warning of impact point and speed parameters as an example to illustrate the algorithm. And here are a few symbols concerned.

W is the external measurement value, J is the warning line, the signal S_I produced within the computer which values 0 or 1, V_k is the launch speed, β_I is the impact point parameter at some moment.

$$S_1\left(W, \beta_c, 10, J\right) = \begin{cases} 1 & \text{if 10 external measurements exceeding the threshold} \\ 0 & \text{otherwise} \end{cases} \tag{5.1}$$

$$S_2\left(W, V_k, 10, J\right) = \begin{cases} 1 & \text{if 10 external speed measurement exceeding the threshold} \\ 0 & \text{otherwise} \end{cases} \tag{5.2}$$

When both S_1 and S_2 are valued 1, the external measurement impact point and velocity parameters of the combined warning system are triggered, that is, $S_W = S_1(W, \beta_c, 10, J) \wedge S_2(W, V_k, 10, J) = 1$.

The safety estimation table (Table 5.1) serves as the reference for safety estimation of the vehicle during launch and flight, as shown in Figure 5.1.

Table 5.1   Safety estimation table.

Category	Sequence	Curvilinear coordinates	Alarming line	Blowing-up line
Trajectory dropping	1	Dropping point	√	√
point by external	2	Range	√	√
measurement	3	Speed	√	√
	4	Dip	√	
	5	Declination	√	
Trajectory dropping	6	Dropping point	√	√
point by remote	7	Range	√	√
measurement	8	Speed	√	√
	9	Dip	√	
	10	Declination	√	
Attitude parameter	11	Pitch angle deviation	√	
by remote	12	Roll angle deviation	√	
measurement	13	Yaw angle deviation	√	
Pressure parameter	14	First-class pressure 1	√	
by remote	15	First-class pressure 2	√	
measurement	16	First-class pressure 3	√	
	17	First-class pressure 4	√	
	18	Second-class pressure	√	
...	...	...	...	...

"√" means the parameter is suitable for estimation of correspondent term.

5.1.3.2 Judgment of Overrun

The overrun parameter (state parameter and impact point parameter) is used to decide whether the vehicle has exceeded the warning line or blowing-up line. The vehicle is deemed within the limit when staying between the two lines and off-limit when being beyond the two lines. Here the warning line of impact point is taken as an example for illustration.

We could find $\lambda_{i-1} < \lambda_i < \lambda_c < \lambda_{i+1}$ that corresponds to λ_i by the longitude and latitude parameter λ_c and β_c of an impact point at one moment, of which $i = 1, 2, \ldots, n$ (theoretical impact point longitude). We get β'_c, the correspondent theoretical impact point longitude of λ_c through $\beta_{i-1}, \beta_i, \beta_{i+1}$ (theoretical impact point latitude) with Lagrange three-point interpolation technique. Use the same technique to calculate $\beta_{c\,up}$ upper-warning limit $\beta_{c\,down}$ and lower-warning limit of λ_c ($\beta_{c\,up}$ donates the upper threshold of β_c while $\beta_{c\,down}$ donates the lower threshold of β_c), the correspondent impact point longitude of upper and lower channel warning values through the impact point longitude of upper and lower channel warning values in the safety database. When $\beta_c \in [\beta_{c\,down}, \beta_{c\,up}]$ is workable, it is deemed that the vehicle stays within the warning lines. Otherwise, it is thought to have exceeded the warning line, and we could calculate the D-value between β_c and the warning line:

$$\Delta\beta_{c\,up} = \beta_{c\,up} - \beta_c$$
$$\Delta\beta_{c\,down} = \beta_c - \beta_{c\,down}$$

(5.3)

The formula above is also suitable for the estimation of the blowing-up line as well as other curvilinear.

5.1.3.3 Selection of Impact Point

When choosing the impact point, we could get three pairs of impact point parameters by instantaneous impact point parameters $X_{ci}, Z_{ci}, \sin\delta_c, \cos\delta_c$ ($i = 1, 2, 3$), among which X_{ci}, Z_{ci} are the launch coordinates and δ_c is the geodetic azimuth projected on the ground by speed vector, when the three pairs of impact point longitude and latitude parameter λ_c, β_c meet the requirements simultaneously. It is deemed that the instantaneous impact point meets the requirements of impact point selection. Otherwise it shall not be selected.

The selection of impact points involves the following steps: First select an impact point. If successful, there is no need for another selection; if not, a second selection is required. When the second try also fails, it means the instantaneous impact point does not meet the requirement of impact point selection. Calculate the selection of impact point, and if the protective circles used to protect cities concerned to deviate from the ellipse of the vehicle's debris distribution, there is no need for further selection. We go through another selection only when it is uncertain whether the protective circles that fail the requirements in the first selection stage intersect or interfere with the debris distribution ellipse.

5.2 Model of Space Launch and Flight Safety Control Parameter Calculation

In the process of space launch, models of vehicle debris separation or malfunction-stricken vehicle falling trajectory as well as impact point area analysis are the key issues that have to be resolved in safety control. We could calculate the falling trajectory and impact point according to booster, first-class launch vehicle, fairing separation parameters in normal functioning conditions based on impact point calculation model. Once any fault occurs in the process of vehicle flight, the system has to calculate the real-time impact point, debris distribution range, and toxic gas spreading area in order to decide key protective district. In addition, the system has to calculate the threatened range, area, and population according to weather and geographical conditions to select the best safety control timing and make clear the hazard level and situation, thus protecting the ground and minimizing losses. Calculations of safety control parameters mainly involve real-time flight trajectory, impact point, safety channel, debris distribution area, toxic gas spreading area, etc.

5.2.1   Safety Channel Calculation

The allowed variation range of trajectory deviation of a launch vehicle in a powered-flight process is also called safety channel. The safety channel is designed based on the impact point boundary, the flight features of malfunction-stricken vehicle, distribution of a protective area as well as error range affecting safety control. According to specific trajectory parameters of the launch vehicle, the safety channel could be divided into three types, namely position, speed, and impact point. In actual circumstance, the three safety channels are drawn on two-dimensional curve chart, which are called real-time position, real-time speed, and real-time impact point safety channel charts respectively. It could be drawn continuously on the chart or demonstrated on the screen. The safety channel serves as a basic reference for estimation of the flight state of launch vehicles. The calculation involves the following steps:

1. Theoretical trajectory interpolation: Calculate the value at a specific time with theoretical trajectory data within the safety control period based on polynomial three-point interpolation technique.
2. Channel deviation calculation: Channel deviation refers to the relevant error between theoretical and actual data in terms of safety control warning parameters and blowing-up channel. The channel deviation at a specific moment is calculated according to the following formula:

$$\delta = \sqrt{\left(k_1\delta_{gr}\right)^2 + \left(k_2\delta_{cl}\right)^2 + \left(\delta_{xs}\right)^2 + \left(\delta_{sy}\right)^2 + \left(\delta_{sm}\right)^2} \tag{5.4}$$

In the equation, δ_{gr} represents the flight deviation of a launch vehicle and δ_{cl} is the measured deviation. When we are calculating the warning channel, we adopt the deviation of high accuracy while using deviation of lower accuracy of the blowing-up channel. The point of δ_{xs} is the demonstrated deviation, δ_{sy} is the time-lag deviation from a delivery system, δ_{sm} is the deviation of a mathematical model, and k_1, k_2 are used to calculate warning and blowing-up channel coefficient. The result of $(\delta_{xs})^2 + (\delta_{sy})^2 + (\delta_{sm})^2$ is constant when calculating a deviation of various channels. C_1 marks the deviation channel of impact point and position parameter, C_2 for speed deviation channel, and C_3 for angel deviation channel.
3. Determination of blowing-up and warning channel: Calculate the blowing-up deviation δ_{bz} and warning line δ_{gj} interval interpolation, and align it with the theoretical trajectory with a

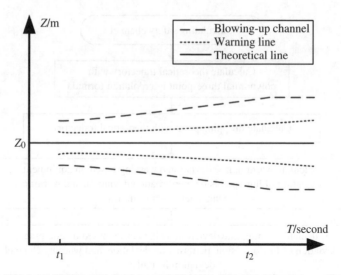

Figure 5.2 Lateral deviation Z safety channel.

polynomial three-point interpolation formula, from which we get the blowing-up and warning channel $GD_{\chi_{bw}}$:

$$GD_{\chi_{bw}} = X \pm \delta_{\chi_{bw}} \tag{5.5}$$

In the formula, "+" means the upper channel, otherwise it is the lower channel, and X is the theoretical trajectory value. The lateral deviation Z safety channel is shown in Figure 5.2, and Figure 5.3 offers the safety channel calculation flowchart.

5.2.2   Multisourced Data Fusion of Launch Vehicle Trajectory Parameters

Data fusion is an information-processing technique targeting system with more than one or one type of sensor. It is also called multisource association, multisource synthesis, and mixing of sensors. However, in most cases, it is called multisource sensor data fusion, or data fusion for short. Data fusion is generally defined as follows: analyze automatically and optimize measurement information of sensors acquired in sequence by computer techniques under certain rules to complete information processing needed by decision-making and task estimation. According to this definition, data fusion takes the sensor as the foundation, multi-source information the target, and coordination and comprehensive treatment the core.

For the trajectory measurement system of launch vehicle, data fusion would estimate the data in a real-time manner, process the data of all or partial measurement equipment under certain rules, and generate the final flight data of the target, which is displayed and controlled by the launch command center. The data fusion into a trajectory measurement system belongs to position data fusion. The section will focus on the features of trajectory measurement system and propose suggestions for optimizing real-time flight data processing capability of trajectory measurement system and its efficiency by data fusion as well as improving data reliability, accuracy, and lowering uncertainties.

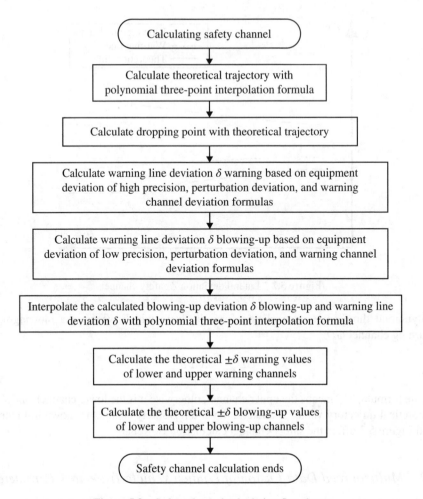

Figure 5.3　Safety channel calculation flowchart.

The trajectory measurement data of the launch vehicle include measurement data of the telemetry system and the external measurement system, such as radars, infrared sensor, photoelectric theodolite, telemetry data receiver (platform guidance data and strap-down guidance data), and GPS measurement, which reflects objectively the features of the measurement target system from different perspectives. The measurement data of the telemetry system enjoys relative stability while that of the external measurement system is prone to interferences.

5.2.2.1　Preprocessing of Telemetry and External Measurement Data

When monitoring the flight of a launch vehicle, the actual signal is often found with singular terms or trend terms coupled with periodic interference and noise interference. In addition, the sensor and converter may also be caught with zero drifts. In order to improve the quality of the data, it is necessary to preprocess the measurement data. The preprocessing stage involves data examination, data processing, data selection, and data smoothing.

Data Validation
In data validation, we first validate the accuracy and smoothness of measurement data to ensure if the signal is flooded by noises and determines its nature. This stage mainly aims at determining the accuracy of signal as well as its smoothness, periodicity, and normality.

Data Processing
In this stage, we would process the data collected in a proper way, including removal of trend terms, suppression of periodic interference signal, and elimination of wild values.

1. Elimination of trend terms: The frequency whose period lasts longer than selected section is called trend term. The trend term emerges, on one hand, from distorted baseline drift, and on the other hand, from disturbances of lower or low frequency compared with the main frequency band. We could get rid of trend terms by the least square method.
2. Suppression of periodic interference: In processing the telemetry speed-variation parameter of a launch vehicle, the frequency spectrum and power spectrum of such parameters might be disturbed by periodic disturbance component from certain frequency and its harmonic wave due to interference from power frequency, frame code, and sampling frequency as well as effects from calibrating signal of the vehicle's records. As it exerts even more influence upon the low-frequency parameters, we have to suppress such interference. We recommend suppressing periodic interference with filter, which, in essence, sets the amplitude of the signal concerned at zero only on the disturbing frequency point without affecting other points. In actual practices, it is suggested that the signal amplitude of small neighborhood around the disturbing frequency point should be set close to zero.
3. Elimination of wild values: Wild value, also called jump point, refers to abnormal hopping point caused by interference from sensor, converter, and the delivery process, rather than normal trip points of the tested target. The wild value produces false harmonic wave components, which would affect the energy estimation of the measurement data by reducing signal-to-noise ratio.

Data Selection
In the measurement of the launch vehicle's external trajectory, the data is often found with abnormal value of serious error. This phenomenon may have many reasons, including facility faults, mistakes in data record and reading, sudden change of surroundings, interferences and misconduct by an operator. The abnormal data observed by the same measurement equipment displays different forms and valves, and the difference is even greater using different measurement equipment. Such abnormal data is called wild value, which proves harmful to the precision to the result of data processing and may distort partially or even fully the results. The most commonly used data selection approach for single measurement element takes advantage of temporal correlation of real signals in the time sequence of single measurement element. Therefore, this approach is highly efficient in eliminating separate wild value but almost helpless in identifying wild value stretch with correlated properties for a long time. And it cannot identify system errors that undergo slow changes.

The data selection approach based on trajectory parameters uses the filtered trajectory parameters $(x, \dot{x}, y, \dot{y}, z, \dot{z})$ for authentic assessment and carries out data systematic error estimation and data selection through back-calculation of the difference between each measurement element and measurement unit. Take the data selection as an example. We are getting X $(t+1)$ of the trajectory parameter $X(t)$ at the present moment with a prediction algorithm. And based on the characteristics of the measurement unit and elements, we get $\tilde{R}(t+1)$, $\tilde{A}(t+1)$, $\tilde{E}(t+1)$, $\tilde{\dot{R}}(t+1)$. And by calculating the difference between the result with the actual measurement $R(t+1)$, $A(t+1)$, $E(t+1)$, $\dot{R}(t+1)$, we get $\Delta R(t+1)$, $\Delta A(t+1)$, $\Delta E(t+1)$, $\Delta \dot{R}(t+1)$. Set up the threshold value Mx to decide if it meets the following formula: $|\Delta x(t+1)| \leq Mx$.

The threshold value could be a fixed value or self-adaptive threshold value.

The data selection approach based on trajectory parameters considers not only the time correlation of measurement elements but also the reality of signals. If all the measurement elements are found with only random errors, this approach is apparently much better than single element data selection in distinguishing wild values. When serious systematic error occurs in the measurement element, it would diverge the trajectory if it takes part in trajectory parameter calculation while elevating threshold values of other measurement elements in order to make them pass the data selection. If the measurement element is not involved in parameter calculations, the residual historic data would display dramatically trend terms, which could offset serious systematic error in a real-time manner.

Temporal and Space Alignment

In data fusion of multisensors, each group of measurement data may have been different sampling intervals and measurement coordinates due to the difference between measurement equipment and its geographical positions. Even internal sensors from the same measurement unit may use completely different sampling intervals and measurement coordinates. For this reason, it is necessary to align the measurement data with the unified time and space coordinates prior to data fusion. The process is divided into two steps, namely coordinate system transformation and systematic error correction. In a number of literatures, the correction of systematic error is also called systematic error registration. It is found that without estimation and correction of systematic error, when each sensor transferred the measurement data of the same target to the unified coordinate system, the target would show multi-tracks due to systematic error in the process of data fusion. This would lead to the failure of data fusion. And after its completion, the precision of the data is still worsened after the fusion.

1. Time alignment: The zero point on the clock of the launch vehicle is usually the moment when the vehicle is launched and recorded in the form of absolute time. The telemetry data outputs data at certain sampling intervals according to the telemetry clock control on the launch vehicle. The data sampling timing of the external measurement system on the ground is recorded in the form of absolute time provided by the ground clock control. However, the sampling intervals differ from the sampling period of the system. Therefore, it is necessary to align external trajectory data working asynchronously with the telemetry data in terms of timing, ensuring that they are integrated at the same time. This step is a primary issue to be resolved before data fusion. At present, the most commonly used algorithms include the least square method, Taylor expansion, Lagrange's interpolation, and neural network fitting method.

2. Space alignment: Space alignment refers to the fusion of measurement data from a different coordinate system in the same coordinate system. Take the launch coordinate system, a unified baseline coordinate system, as an example. Now we assume that the measurement data of a measurement equipment is:

$$V_i = \begin{pmatrix} A_i \\ E_i \\ R_i \end{pmatrix}, \quad i = 1,\dots,n$$

In the formula, A_i, E_i, R_i represent the azimuth, elevation, and distance measured by equipment i in its measurement system (polar coordinate system). Now transform it into the rectangular coordinate system:

$$
\begin{aligned}
x_i^c &= R_i \cos\left(E_i\right)\cos\left(A_i\right) \\
y_i^c &= R_i \sin\left(E_i\right) \\
z_i^c &= R_i \cos\left(E_i\right)\sin\left(A_i\right)
\end{aligned}
\tag{5.6}
$$

Then transform the measurement coordinate system into a launch coordinate system. Set the launch coordinate of the transformed measurement equipment i as $\left(x_i^f, y_i^f, z_i^f\right)$ and the measurement system coordinates $\left(x_i^c, y_i^c, z_i^c\right)$, then the transformation correlation is represented by:

$$
\begin{pmatrix} x_i^f \\ y_i^f \\ z_i^f \end{pmatrix} = M \cdot \begin{pmatrix} x_i^c \\ y_i^c \\ z_i^c \end{pmatrix} + G_i^0
\tag{5.7}
$$

In the formula, M_i is the rotation transformation matrix from the equipment's original coordinate system into the launch coordinate system and G_i^0 is the translation transformation matrix.

5.2.2.2 Telemetry and External Measurement Multisource Data Fusion Model

The distributed multisensor data fusion system could lower the demands for the sensor capability of the system and communication bandwidth, thus reducing the cost of the system; in addition, due to its structural advantages, the distributed fusion system displays reliability, vigor, and fast decision-making, being widely used in engineering technologies. In the distributed multisensor target tracking system, each sensor, before its measurement data is delivered into the fusion center, would generate a partial trajectory estimation in its data processor, and then send the processed estimation data to the fusion center where coordinate transformation, time alignment, or corrections are accomplished, and finally the center produces an overall estimation based on the trajectory estimation at each node. How to effectively integrate partial trajectory estimation and make full use of the redundant and complement information form the key issues of distributed flight path fusion. The dynamic distributed multilevel data fusion proposes the following: Calculate the level of support between the estimated trajectory measurement values from each single measurement unit, select two

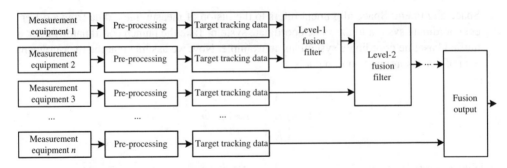

Figure 5.4   Model of dynamic distributed multilevel data fusion structure.

groups of the higher support level for data fusion, calculate the fusion value and the estimated residual value, and integrate the estimated value of the higher support level again. And this procedure is repeated for a few times. Through the approach of dynamic selection of fusion sequence of sensors, the multilevel fusion of measurement data from multisensors is achieved. The structure is shown in Figure 5.4.

Dynamic Examination of Data Differences

From the perspective of information fusion the closer the difference between measurement data of each sensor, the higher the support level of sensor measurement is. If the fusion of two sensors of a high support level could reflect better the measured parameter, such fusion will surely reduce uncertainties in the process of measurement. If the two do not match and one of them has to blunder error, then fusion of the two would even worsen the result, making fusion impossible. Therefore, in the process of data fusion, we need to conduct a consistency check of measurement data from multisources, which serves as references for dynamic fusion sequence.

As to the issue of a support level between trajectory data, we take integral difference between two trajectory zones within a certain sampling interval as the reference, which is defined as:

$$d_{ij} = \int_{t}^{t+\Delta t} \sqrt{\left(x_i + \dot{x}_i t - x_j - \dot{x}_j t\right)^2 + \left(y_i + \dot{y}_i t - y_j - \dot{y}_j t\right)^2 + \left(z_i + \dot{z}_i t - z_j - \dot{z}_j t\right)^2} \, dt \qquad (5.8)$$

In the formula, x_i, y_i, z_i represent the coordinate of sensor i in the rectangular coordinate system, $\dot{x}_i, \dot{y}_i, \dot{z}_i$ represent the target speed of sensor i in the rectangular coordinate system, and d_{ij} is the trajectory integral between sensor i and sensor j.

Extend formula (5.8), and we get:

$$d_{ij} = \int_{t}^{t+\Delta t} \sqrt{at^2 + bt + c} \, dt \qquad (5.9)$$

where
$$\begin{cases} a = \left(\dot{x}_i - \dot{x}_j\right)^2 + \left(\dot{y}_i - \dot{y}_j\right)^2 + \left(\dot{z}_i - \dot{z}_j\right)^2 \\ b = 2\left[\left(x_i - x_j\right)\left(\dot{x}_i - \dot{x}_j\right) + \left(y_i - y_j\right)\left(\dot{y}_i - \dot{y}_j\right) + \left(z_i - z_j\right)\left(\dot{z}_i - \dot{z}_j\right)\right] \\ c = \left(x_i - x_j\right)^2 + \left(y_i - y_j\right)^2 + \left(z_i - z_j\right)^2 \end{cases}$$

Here, the result of $at^2 + bt + c > 0$, and we get

$$d_{ij} = \int_t^{t+\Delta t} \sqrt{at^2 + bt + c}\, dt$$

$$= \frac{2at+b}{4a}\sqrt{a\Delta t^2 + b\Delta t + c} + \frac{4ac-b^3}{8a^{3/2}} \ln\left|2at + b + 2\sqrt{a}\sqrt{a\Delta t^2 + b\Delta t + c}\right| \qquad (5.10)$$

$$- \frac{b}{4a}\sqrt{c} - \frac{4ac-b^3}{8a^{3/2}} \ln\left|b + 2\sqrt{ac}\right|$$

As the formula above is quite complex, we, therefore, propose a simplified formula, which involves integral and extraction so as to prove their similarity. The simplified formula is shown as below:

$$d_{ij} = \sqrt{\int_t^{t+\Delta t}\left(at^2 + bt + c\right)dt}$$

$$= \sqrt{a\left(t^2\Delta t + t\Delta t^2\right) + \frac{1}{2}b\left(2t\Delta t + \Delta t^2\right) + c\Delta t} \qquad (5.11)$$

It is concluded that the smaller the d_{ij}, the higher the support level between sensor i and sensor j, which in turn, reflects the state variation of the measured data more consistently, thus reducing uncertainties in the process of measurement. On the contrary, the lower the support level is between the two, the worse the result of their fusion will be. Therefore, d_{ij} could be used as a parameter in deciding the consistency between sensor i and sensor j.

For the case that involves N sensors, take $d_{ij}(i, j = 1,2,\ldots,N)$ as the elements, then the distance matrix is defined as:

$$D = \begin{bmatrix} d_{11} & d_{12} & \cdots & d_{1N} \\ d_{21} & d_{22} & \cdots & d_{2N} \\ \vdots & \vdots & \ddots & \vdots \\ d_{N1} & d_{N2} & \cdots & d_{NN} \end{bmatrix} \qquad (5.12)$$

The search for the closest measurement element to the trajectory distance in the algorithm is carried out as follows:

1. Set the correspondent threshold ε.
2. Calculate the statistical distance $d_{ij}(i, j = 1,2,\ldots,N)$ between each pair of sensors, in which N represents the total number according to formula (5.11) and from the distance matrix D according to formula (5.12).
3. Eliminate elements on the diagonal line of the lower triangular matrix in distance matrix D and search the pair of sensors p and q, the distance of which is the shortest and does not exceed ε, that is,

$$d_{pq} = \min_{\substack{i,j=1,2,\ldots,N \\ i>j}}\left\{d_{ij}\right\} \quad \text{and} \quad d_{pq} \leq \varepsilon \qquad (5.13)$$

4. Integrate the measurement value of sensor p and q according to the fusion algorithm, the result of which is used as the new measurement value of sensor p. At the same time, delete sensor q among the sensors that take part in the current data fusion, subtract the number of sensor by 1 whose number exceeds q, and maintain the measurement value. By this way, we get a new sensor group that participates in data fusion and the sensors of which are subtracted by 1. After this step, recalculate the statistical distance between each sensor and from a new distance matrix D.
5. Repeat steps (3) and (4) until the sensor pair whose distance is the shortest and does not exceed ε is found. At the moment, the fusion result of the sensor group taking part in current data fusion is the result of final multisensor measurement value consistency fusion.

From the definition of statistical distance in formula (5.8), it is clear that no matter the measurement value of sensor has blunder error, the statistical distance from the sensor measurement value to itself is zero, that is, the sensor is always the most supportive to itself. Therefore, it is impossible to measure the consistency of sensor measurement value by the statistical distance between sensor measurement values to itself. We could only use the statistical distance between different sensor measurement values to examine the consistency of sensor measurement values. For this reason, it is necessary to eliminate elements on the diagonal line of the lower triangular matrix in step (3).

From the aforementioned five steps, we could determine the sequence of the fusion of trajectory estimation values of each single measurement unit to achieve dynamic multilevel data fusion.

5.2.2.3   Telemetry and External Measurement Multisource Data Fusion Algorithm

For the multisensor measurement system, the aim of position-level data fusion in the process of real-time tracking is to estimate data efficiency, integrate the data according to certain rules, and then send the data back to the control system, which will then take part in the control process. To ensure the efficiency of the data delivered, the fusion rules played a very important role in the process. And the adjustment of a measurement data ratio through weighting efficient is highly effective for this aspect and the determination of a weighting factor lies at the core of the approach.

In the process of dynamic multilevel fusion, we have to find out two pairs of measurement data of high supportive level for fusion in the first place, and calculate the supportive level between the fusion result and other values, which will be followed by further pairwise fusion. For this reason, the section will first study the fusion algorithm of pairwise sensors and then propose a recursion formula for multisensors.

Matrix Weighting Optimal Fusion Estimation Algorithm of Pairwise Sensors
The fusion of weighted data refers to the process that multisensors first measure the data of the same characteristics' parameter in an environment, then weight each sensor according to certain rule while considering partial estimation of each sensor, and finally get an optimal overall estimation value by weighting all the partial estimation. The multisensor data fusion

aims at improving the precision of the target estimation. And it is necessary to consider the effect of variance of sensor on weight fusion during fusion.

Set the state vector $X \in R^n$, and based on the observation of two subsystems, we get two unbiased estimation, that is, $\hat{X}_1$ and $\hat{X}_2$, respectively:

$$E\hat{X}_1 = EX, \quad E\hat{X}_2 = EX \tag{5.14}$$

In the formula, E is the mean value. The estimation error is calculated by $\tilde{X}_1 = X_1 - \hat{X}_1$ and $\tilde{X}_2 = X_2 - \hat{X}_2$, respectively.

Set the estimation error variance matrix and the covariance matrix, respectively, as $P_1 = E\left[\tilde{X}_1 \tilde{X}_1^T\right], P_2 = E\left[\tilde{X}_2 \tilde{X}_2^T\right], P_{12} = E\left[\tilde{X}_1 \tilde{X}_2^T\right], P_{21} = E\left[\tilde{X}_1 \tilde{X}_2^T\right] = P_{12}^T$.

Set the fusion estimation of X based on estimation value $\hat{X}_1$ and $\hat{X}_2$ as:

$$\hat{X}_0 = A_1 \hat{X}_1 + A_2 \hat{X}_2 \tag{5.15}$$

In the formula, A_1 and A_2 is the $n \times n$ dimension weighted matrix.

The problem is how to minimize error of mean square of components by weighted matrix A_1 and A_2

$$J = \mathrm{tr}P_0, \quad P_0 = E\left[\tilde{X}_0 \tilde{X}_0^T\right], \quad \tilde{X}_0 = X - \hat{X}_0 \tag{5.16}$$

In the formula, the symbol "tr" means the trace of the matrix.

According to formula (5.15) and the unbiased characteristics of estimation value $\hat{X}_1$, $\hat{X}_2$, $\hat{X}_0$, it is concluded that $EX = E\hat{X}_0 = A_1 E\hat{X}_1 + A_2 E\hat{X}_2 = (A_1 + A_2)EX$.

From the formula above, we get:

$$A_1 + A_2 = I_n \tag{5.17}$$

And

$$\tilde{X}_0 = A_1 \tilde{X}_1 + A_2 \tilde{X}_2 \tag{5.18}$$

From formula (5.17), we obtain $A_2 = I_n - A_1$, and by introducing it into formula (5.18), we get $\tilde{X}_0 = A_1 \tilde{X}_1 + (I_n - A_1)\tilde{X}_2$.

From the above formula, we get:

$$
\begin{aligned}
P_0 &= E\left[\tilde{X}_0 \tilde{X}_0^T\right] \\
&= A_1 P_1 A_1^T + (I_n - A_1) P_2 (I_n - A_1)^T + (I_n - A_1) P_{21} A_1^T + A_1 P_{12} (I_n - A_1)^T \\
&= A_1 (P_1 + P_2 - P_{12} - P_{21}) A_1^T + P_2 + A_1 (P_{12} - P_2) + (P_{21} - P_2) A_1^T
\end{aligned}
$$

And according to the derivation formula

$$\frac{\partial}{\partial Y}\mathrm{tr}(YB) = B^T, \quad \frac{\partial}{\partial Y}\mathrm{tr}(BY)^T = B, \quad \frac{\partial}{\partial Y}\mathrm{tr}(YBY^T) = Y(B^T + B)$$

By setting $\partial \mathrm{tr} P_0 / \partial A_1 = 0$, we get: $2A_1(P_1 + P_2 - P_{12} - P_{21}) + 2(P_{21} - P_2) = 0$ and we have:

$$
\begin{aligned}
A_1 &= (P_2 - P_{21})(P_1 + P_2 - P_{12} - P_{21})^{-1} \\
A_2 &= I_n - A_1 = \left[(P_1 + P_2 - P_{12} - P_{21}) - (P_2 - P_{21}) \right](P_1 + P_2 - P_{12} - P_{21})^{-1} \qquad (5.19) \\
&= (P_1 - P_{12})(P_1 + P_2 - P_{12} - P_{21})^{-1}
\end{aligned}
$$

Now by introducing formula (5.19) into formula (5.15), we get the minimum variance optimal fusion estimation:

$$
\hat{X}_0 = (P_2 - P_{21})(P_1 + P_2 - P_{12} - P_{21})^{-1} \hat{X}_1 + (P_1 - P_{12})(P_1 + P_2 - P_{12} - P_{21})^{-1} \hat{X}_2 \qquad (5.20)
$$

By introducing expression formula of A_1 into formula (5.16), we get the minimum error variance matrix:

$$
P_0 = P_2 - (P_2 - P_{21})(P_1 + P_2 - P_{12} - P_{21})^{-1} (P_2 - P_{21})^T \qquad (5.21)
$$

And $\mathrm{tr} P_0 \leq \mathrm{tr} P_1$, $\mathrm{tr} P_0 \leq \mathrm{tr} P_2$.

Pairwise Sensors Scalar-Weighted Optimal Fusion Estimation Algorithm

The above formula of pairwise sensors scalar-weighted optimal fusion estimation algorithm for the purpose of optimal data fusion estimation (5.20) requires calculation of an inverse matrix of $n \times n$ dimension matrix $(P_1 + P_2 - P_{12} - P_{21})$ in preparation for the calculation of a weighted matrix. However, this would affect real-time application to some degree. In order to reduce the burden of online calculation and facilitate real-time application, here we propose a simplified scalar-weighted minimum variance optimal fusion estimation formula:

$$
\hat{X}_0 = a_1 \hat{X}_1 + a_2 \hat{X}_2
$$

In the formula, $\hat{X}_0, \hat{X}_1, \hat{X}_2$, and $X \in R^n$; a_1 and a_2 represent the scalar weighting coefficient. The issue is how to choose a_1 and a_2 to minimize the sum of scalar estimation error and mean square error:

$$
J = \mathrm{tr}\, P_0, \quad P_0 = E\left[\tilde{X}_0 \tilde{X}_0^T \right], \quad \tilde{X}_0 = X - \hat{X}_0 \qquad (5.22)
$$

From the unbiased characteristics of estimation value $\hat{X}_1, \hat{X}_2$, and $\hat{X}_0$, we get:

$$
a_1 + a_2 = 1,
$$

and

$$
\tilde{X}_0 = a_1 \tilde{X}_1 + a_2 \tilde{X}_2 \qquad (5.23)
$$

By introducing $a_1 = 1 - a_2$ into the above formula, we get:

$$P_0 = E\left[\tilde{X}_0 \tilde{X}_0^T\right] = (1 - a_2)^2 P_1 + a_2^2 P_2 + (1 - a_2)a_2 P_{12} + a_2(1 - a_2)P_{21} \tag{5.24}$$

And we get $\mathrm{tr}P_0 = (1 - a_2)^2 \mathrm{tr}P_1 + a_2^2 \mathrm{tr}P_2 + 2(1 - a_2)a_2 \mathrm{tr}P_{12}$. Set $\partial \mathrm{tr}P_0 / \partial a_2 = 0$, so that $-2(1 - a_2)\mathrm{tr}P_1 + 2a_2 \mathrm{tr}P_2 + 2(1 - 2a_2)\mathrm{tr}P_{12} = 0$

So we get:

$$a_2 = \frac{\mathrm{tr}P_1 - \mathrm{tr}P_{12}}{\mathrm{tr}P_1 + \mathrm{tr}P_2 - 2\mathrm{tr}P_{12}} \tag{5.25}$$

From the relation $a_1 = 1 - a_2$, we get:

$$a_1 = \frac{\mathrm{tr}P_2 - \mathrm{tr}P_{12}}{\mathrm{tr}P_1 + \mathrm{tr}P_2 - 2\mathrm{tr}P_{12}} \tag{5.26}$$

By introducing formula (5.27) and (5.28) into formula (5.25), we get the minimum error variance matrix:

$$P_0 = a_1^2 P_1 + a_2^2 P_2 + 2a_1 a_2 P_{12} \tag{5.27}$$

The sum of minimum component estimation error and mean square is: $\mathrm{tr}P_0 = \mathrm{tr}\left[a_1^2 P_1 + a_2^2 P_2 + 2a_1 a_2 P_{12}\right]$
Set $a_1 = 1, a_2 = 0; a_1 = 0, a_2 = 1$ in formula (5.16), we get $\mathrm{tr}P_0 \leq \mathrm{tr}P_1, \mathrm{tr}P_0 \leq \mathrm{tr}P_2$.

Targeting the system which is observed by multisensors simultaneously, the section proposes the optimal information fusion principle featuring scalar-weighted linear minimum variance to deal with scalar-weighted optimal information fusion estimation related to partial subsystem estimation error. Based on the information fusion principle, the section proposes scalar-weighted optimal information fusion steady-state Kalman filter, which gets weighted coefficient through just one fusion when the subsystem of each single sensor reaches a steady state to avoid calculations of the weighted coefficient that have to be conducted all the time and reduce calculation burdens.

Information Fusion Algorithm Based on Diagonal Matrix Weighting

We have studied information fusion algorithm based on matrix weighting. As the algorithm requires to calculate Riccati equation, optimal weighting matrix, and optimal filter gain matrix online, it will inevitably produce huge calculation burden in face of high-state dimension, making it inefficient in real-time application. To solve this problem, this section proposes a scalar weighting information fusion algorithm, which is convenient but leads to loss of partial fusion precision.

As the weighting fusion estimation based on diagonal matrix equals that of the state scalar, the decoupling information fusion Kalman filter is achieved, that is, the component i of fusion state estimation is the fusion estimation of partial estimation component I based on scalar weighting, or the fusion estimation of the same physical parameter based on scalar weighting, making it highly applicable.

Based on the "pairwise sensors matrix weighting optimal fusion estimation algorithm," set the state variant X and its unbiased estimation $\hat{X}_1$ and $\hat{X}_2$, and fusion estimation $\hat{X}_0$ as:

$$
X = \begin{bmatrix} x_1 \\ \vdots \\ x_n \end{bmatrix}, \quad
\hat{X}_1 = \begin{bmatrix} \hat{x}_{11} \\ \vdots \\ \hat{x}_{1n} \end{bmatrix}, \quad
\hat{X}_2 = \begin{bmatrix} \hat{x}_{21} \\ \vdots \\ \hat{x}_{2n} \end{bmatrix}, \quad
\hat{X}_0 = \begin{bmatrix} \hat{x}_{01} \\ \vdots \\ \hat{x}_{0n} \end{bmatrix}
\tag{5.28}
$$

To reduce the calculation burdens of matrix weighting fusion algorithm and improve the precision of scalar weighting fusion algorithm, we propose the linear minimum variance optimal fusion estimation based on diagonal matrix weighting. Set the fusion estimation value as:

$$
\begin{bmatrix} \hat{x}_{01} \\ \hat{x}_{02} \\ \vdots \\ \hat{x}_{0n} \end{bmatrix} =
\begin{bmatrix} a_{11} & 0 & \cdots & 0 \\ 0 & a_{12} & \cdots & 0 \\ \vdots & \vdots & \ddots & \vdots \\ 0 & 0 & \cdots & a_{1n} \end{bmatrix}
\begin{bmatrix} \hat{x}_{11} \\ \hat{x}_{12} \\ \vdots \\ \hat{x}_{1n} \end{bmatrix} +
\begin{bmatrix} a_{21} & 0 & \cdots & 0 \\ 0 & a_{22} & \cdots & 0 \\ \vdots & \vdots & \ddots & \vdots \\ 0 & 0 & \cdots & a_{2n} \end{bmatrix}
\begin{bmatrix} \hat{x}_{21} \\ \hat{x}_{22} \\ \vdots \\ \hat{x}_{2n} \end{bmatrix}
\tag{5.29}
$$

Now the issue is how to choose a diagonal matrix.

$$
A_1 = \mathrm{diag}\left(a_{11}, a_{12}, \ldots, a_{1n}\right), \quad A_2 = \mathrm{diag}\left(a_{21}, a_{22}, \ldots, a_{2n}\right)
\tag{5.30}
$$

$$
J = \mathrm{tr}P_0, \quad P_0 = E\left[\tilde{X}_0 \tilde{X}_0^T\right], \quad \tilde{X}_0 = X - \hat{X}_0
\tag{5.31}
$$

Set the scalar estimation error as:

$$
\tilde{x}_{0i} = x_i - \hat{x}_{0i}, \quad \tilde{x}_{1i} = x_i - \hat{x}_{1i}, \quad \tilde{x}_{2i} = x_i - \hat{x}_{2i}
\tag{5.32}
$$

According to formula (5.29), we get:

$$
\hat{x}_{0i} = a_{1i}\hat{x}_{1i} + a_{2i}\hat{x}_{2i}, \quad i = 1, 2, \ldots, n
\tag{5.33}
$$

From the unbiased characteristic of estimation value, we get the following relation:

$$
a_{1i} + a_{2i} = 1, \quad i = 1, 2, \ldots, n
\tag{5.34}
$$

Then $x_i = a_{1i}x_{1i} + a_{2i}x_{2i}$, and by subtracting formula (5.27) from it, we get:

$$
\tilde{x}_{0i} = a_{1i}\tilde{x}_{1i} + a_{2i}\tilde{x}_{2i}, \quad i = 1, 2, \ldots, n
$$

So the capability parameter J is:

$$
J = \sum_{i=1}^{n} E\left(\tilde{x}_{0i}^2\right) = \sum_{i=1}^{n} E\left(a_{1i}\tilde{x}_{1i} + a_{2i}\tilde{x}_{2i}\right)^2
\tag{5.35}
$$

Set the elements of P_1, P_2, and P_{12} at line i and row j as P_1^{ij}, P_2^{ij}, and P_{12}^{ij}, that is, $P_1 = \left(P_1^{ij} \right)$, $P_2 = \left(P_2^{ij} \right)$, and $P_{12} = \left(P_{12}^{ij} \right)$

Then J is

$$J = \sum_{i=1}^{n} \left(a_{1i}^2 P_1^{ij} + a_{2i}^2 P_2^{ij} + 2a_{1i} a_{2i} P_{12}^{ij} \right) \tag{5.36}$$

Under the constraint condition formula (5.34), the minimization of J equals to the scalar weighted optimal fusion estimation by n, that is,

$$\begin{cases} \min_{a_{1i}, a_{2i}} J_i = a_{1i}^2 P_1^{ij} + a_{2i}^2 P_2^{ij} + 2a_{1i} a_{2i} P_{12}^{ij} \\ a_{1i} + a_{2i} = 1 \end{cases}, \quad i = 1, 2, \ldots, n \tag{5.37}$$

According to diagonal weighted optimal fusion algorithm, the following accounts are described:

Set the two unbiased estimation of random vector $X \in R^n$ as $\hat{X}_1$ and $\hat{X}_2$, its estimation error variance matrix P_1 and P_2, and the estimation error covariance matrix P_{12}, then the diagonal matrix weighted linear minimum variance optimal fusion estimation is as follows:

$$\hat{x}_{0i} = a_{1i} \hat{x}_{1i} + a_{2i} \hat{x}_{2i}, \quad i = 1, 2, \ldots, n \tag{5.38}$$

$$a_{1i} = \frac{P_2^{ij} - P_{12}^{ij}}{P_1^{ij} + P_2^{ij} - 2P_{12}^{ij}}, \quad a_{2i} = \frac{P_1^{ij} - P_{12}^{ij}}{P_1^{ij} + P_2^{ij} - 2P_{12}^{ij}} \tag{5.39}$$

or equally

$$\hat{X}_0 = A_1 \hat{X}_1 + A_2 \hat{X}_2 \tag{5.40}$$

where $\hat{X}_0 = \left(\hat{x}_{01}, \ldots, \hat{x}_{0n} \right)^T$, and

$$A_1 = \text{diag} \left(a_{11}, a_{12}, \ldots, a_{1n} \right), \quad A_2 = \text{diag} \left(a_{21}, a_{22}, \ldots, a_{2n} \right)$$

The component minimum estimation error variance

$$P_{0i} = \frac{P_1^{ij} P_2^{ij} - \left(P_{12}^{ij} \right)^2}{P_1^{ij} + P_2^{ij} - 2P_{12}^{ij}}, \quad i = 1, 2, \ldots, n \tag{5.41}$$

The sum of component minimum estimation error variance is:

$$J = \text{tr} P_0 = P_{01} + P_{02} + \cdots + P_{0n}$$

and

$$\text{tr} P_0 \leq \text{tr} P_1, \quad \text{tr} P_0 \leq \text{tr} P_2$$

Especially when $\mathrm{tr}P_0 \leq \mathrm{tr}P_0$ offered by formula (5.40), that is, the precision of diagonal matrices optimal fusion estimation algorithm is higher than that of the scalar-weighted optimal fusion algorithm.

Through comparisons of the estimation precision of the three optimal fusion estimation algorithm above in the condition of linear minimum variance, it is found that the matrix-weighted fusion algorithm enjoys the highest precision, which is followed by diagonal matrix-weighted fusion algorithm and scalar weighted fusion algorithm. This is because the diagonal weight is an exception of matrix weighting while scalar weighting is an exception of diagonal weighting. The scalar-weighted data fusion has fewer calculation tasks but would lose certain fusion precision despite its high applicability. The diagonal matrix-weighted data fusion filter, compared with the matrix-weighted data fusion filter, has much fewer calculation tasks while achieving precision between the matrix-weighted and the scalar-weighted data fusion algorithms, making it a highly efficient algorithm for data fusion filter in real-time application.

Recursive Algorithm of Multisensors Based on Diagonal Matrix-Weighted Fusion

Set the random variant $X \in R^n$, and based on a total of l sensors, it is calculated that a total number of l unbiased estimation is $\hat{X}_1, \hat{X}_2, ..., \hat{X}_l$, that is,

$$E\hat{X}_i = EX, \quad i = 1, 2, ..., l \tag{5.42}$$

The problem is the fusion estimation of X:

$$\hat{X}_0 = A_1\hat{X}_1 + A_2\hat{X}_2 + \cdots + A_l\hat{X}_l \tag{5.43}$$

It is unbiased, that is, $E\hat{X}_0 = EX$, and A_i is weighted matrix. It is required that $\hat{X}_0$ is the optimal under certain linear minimum variance. Especially when $A_i = a_i I_n$, it is the scalar-weighted fusion, and when set as a diagonal matrix, it is a diagonal-weighted fusion.

The optimal recursive fusion estimation of a total number of l sensors in nature is the optimal data fusion of several pairs of pairwise sensors. To make it easy to understand, we will take the optimal recursive fusion algorithm of three sensors below as example to illustrate the diagonal-weighted recursive fusion estimation of the total number of l sensors.

Under the previously stated conditions, the diagonal-weighted trisensors optimal data fusion estimation involves the fusion estimation of two sensors as shown below:

Step 1: Fusion estimation value of estimation values $\hat{X}_1$ and $\hat{X}_2$

$$\hat{X}_4 = A_1\hat{X}_1 + A_2\hat{X}_2 \tag{5.44}$$

$$A_i = \mathrm{diag}(a_{i1}, a_{i2}, ..., a_{in}), \quad i = 1, 2$$

$$a_{1i} = \frac{P_2^{ij} - P_{12}^{ij}}{P_1^{ij} + P_2^{ij} - 2P_{12}^{ij}} \tag{5.45}$$

$$P_4 = A_1 P_1 A_1^T + A_2 P_2 A_2^T + A_1 P_{12} A_2^T + A_2 P_{21} A_1^T \tag{5.46}$$

Step 2: Fusion estimation value of estimation values $\hat{X}_4$ and $\hat{X}_3$

$$\hat{X}_0 = A_3 \hat{X}_3 + A_4 \hat{X}_4 \tag{5.47}$$

$$A_i = \text{diag}(a_{i1}, a_{i2}, \ldots, a_{in}), \quad i = 3,4 \tag{5.48}$$

$$a_{3i} = \frac{P_4^{ij} - P_{34}^{ij}}{P_3^{ij} + P_4^{ij} - 2P_{34}^{ij}}, \quad a_{4i} = \frac{P_3^{ij} - P_{34}^{ij}}{P_3^{ij} + P_4^{ij} - 2P_{34}^{ij}} \tag{5.49}$$

$$P_0 = A_3 P_3 A_3^T + A_4 P_4 A_4^T + A_3 P_{34} A_4^T + A_4 P_{43} A_3^T \tag{5.50}$$

where $P_{34} = E\left[\tilde{X}_3 \tilde{X}_4^T\right] = E\left[\tilde{X}_3 \left(A_1 \tilde{X}_1 + A_2 \tilde{X}_2\right)^T\right]$ is

$$P_{34} = P_{31} A_1^T + P_{32} A_2^T \tag{5.51}$$

and

$$\text{tr}P_0 \leq \text{tr}P_i, \quad i = 1,2,3 \tag{5.52}$$

5.2.2.4 Telemetry and External Measurement Fusion Simulation for Launch Vehicle Flight

Figures 5.5 and 5.6 offer the final simulation results of data fusion, including telemetry and external measurement and the final fusion results during a certain period. The figures show that the external measurement trajectory is disturbed by the equipment itself and the surroundings and is improved through telemetry and external measurement fusion. From the integration of x, y, z change rate (the speed component of a launch vehicle), we find that the fusion trajectory parameter enjoys better smoothness. After comparing the final integrated trajectory and the external measurement-integrated trajectory with the GPS trajectory, we draw an accumulative error quadratic sum comparison figure, as shown in Figures 5.7 and 5.8.

Figure 5.8 shows that the final fusion trajectory enjoys higher precision than the trajectory that goes through only external measurement integration after the fusion processing. It offers effective and reliable evidence for the estimation of information source in the safety control of the launch vehicle, thus laying a foundation for the development of intelligent safety controls supplementary decision system.

5.2.3 Impact Point Calculation

Impact point calculation is implemented in the spacecraft launch, a real-time calculation of the launch vehicles' location in the primary and secondary flight when it falls on the ground in case of launch failure, providing the basis for safety control. Impact point calculation and safety control strategy are closely related and is the main task of the security design of a control system of a launch vehicle.

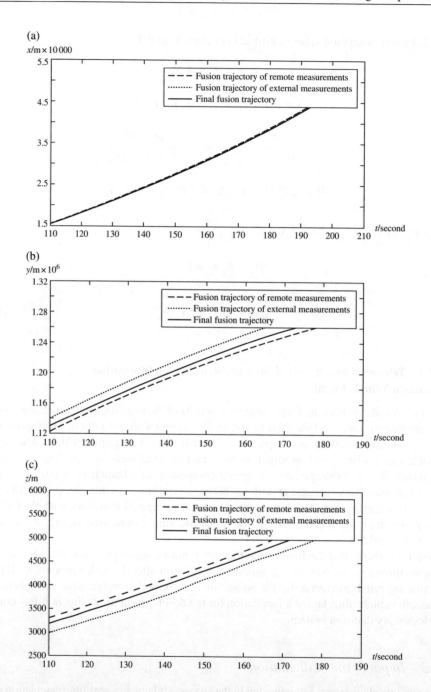

Figure 5.5　External measurement, remote measurement x, y, z, and the final fusion result: (a) integrated x by external and remote measurement, (b) Integrated y by external and remote measurement, and (c) Integrated z by external and remote measurement.

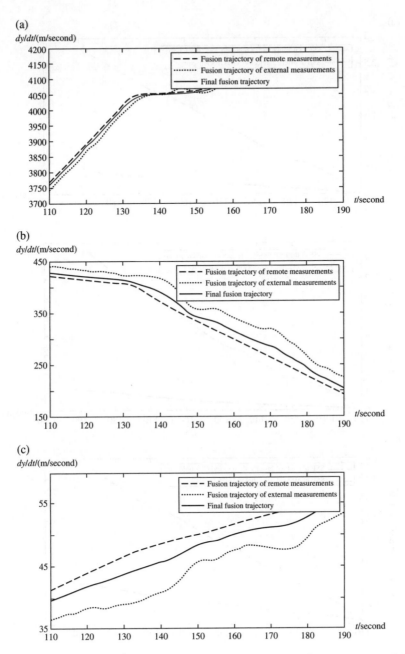

Figure 5.6 External measurement, remote measurement of the change rate of *x, y, z,* and the final fusion result: (a) fusion result of *x* change rate by external and remote measurement, (b) fusion result of *y* change rate by external and remote measurement, and (c) fusion result of *z* change rate by external and remote measurement.

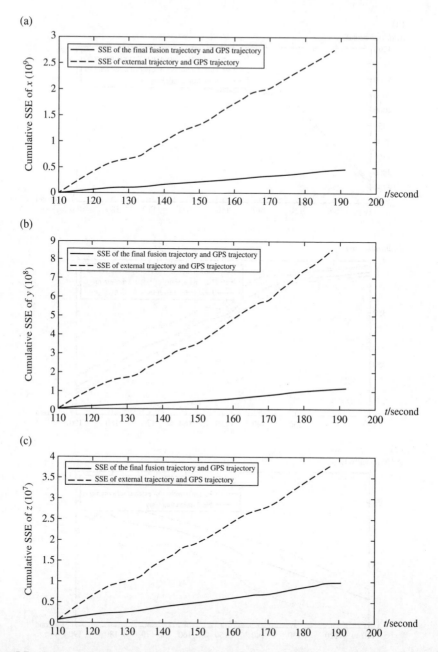

Figure 5.7 Comparison of x, y, z accumulative error quadratic sum: (a) x accumulative error quadratic sum, (b) y accumulative error quadratic sum, and (c) z accumulative error quadratic sum.

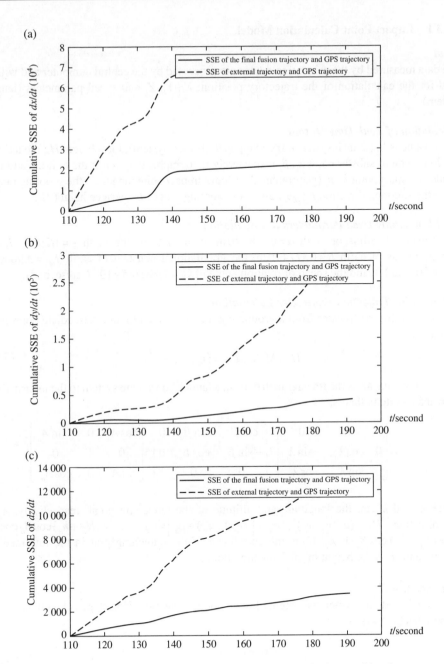

Figure 5.8 Comparison of x, y, z change rate accumulative error quadratic sum: (a) x change rate accumulative error quadratic sum, (b) y change rate accumulative error quadratic sum, and (c) z change rate accumulative error quadratic sum.

5.2.3.1 Impact Point Calculation Model

Input
The data measured by the measuring device are processed by the central computer and will be used for the calculation of the trajectory position: X_k, Y_k, Z_k and speed parameters (launch system) v_{xk}, v_{yk}, v_{zk}.

Calculation of Real-Time Output
Speed value is V_k (launch system); speed dip is θ_k (launch system); local height H_k; the latitude and longitude of substar point λ_k, β_k respectively (geographical system); the latitude and longitude of impact point λ_c, β_c (geographical system); impact point ranges L_c; the shooting ranges along the shooting direction L_x; yaw angle of shooting ranges L_z; perigee height H_p.

Most Commonly Used Parameters in Calculation
The equatorial radius of earth $a = 6378140\,\text{m}$, oblateness of the earth $e = 1/296.257$, the average radius of earth $R = 6371110\,\text{m}$, the gravitational constant of earth $f_M = 3.98600 \times 10^{14}\,(\text{m}^3/\text{second}^2)$, and the speed of rotation angle of $\Omega = 7.292115 \times 10^{-5}\,(\text{rad/second})$.

Formulas for Trajectory Parameter Calculation
The formula for the transfer from a ground coordinate system to an intermediate coordinate system is:

$$U = M_a X + U_a = \left(u_k, v_k, w_k \right)^T \tag{5.53}$$

In the formula, M_a is the transfer matrix from a launch coordinate system to the intermediate coordinate system, that is,

$$M_a = \begin{bmatrix} 1 & 0 & 0 \\ 0 & \cos\lambda_a & -\sin\lambda_a \\ 0 & \sin\lambda_a & \cos\lambda_a \end{bmatrix} * \begin{bmatrix} \cos\beta_a & \sin\beta_a & 0 \\ -\sin\beta_a & \cos\beta_a & 0 \\ 0 & 0 & 1 \end{bmatrix} * \begin{bmatrix} \cos A_a & 0 & -\sin A_a \\ 0 & 1 & 0 \\ \sin A_a & 0 & \cos A_a \end{bmatrix}$$

where λ_a and β_a are the longitude and altitude of the launching point, respectively, A_a is the direction, $U_a = (u_a, v_a, w_a)^T$, $u_a = \left[N_a(1 - e^2) + h_a \right]\sin\beta_a$, $v_a = (N_a + h_a)\cos\beta_a \cos\lambda_a$, $w_a = (N_a + h_a)\cos\beta_a \sin\lambda_a$, U_a is the coordinate of the launching pint in the intermediate system, and H_a is the height of the launching point.

Parabola Model
The model is adopted when the height of local point in the ascending stage is less than 30 km. In the launching system:

$$L_x = x_k + v_{xk} \times t$$
$$L_z = z_k + v_{zk} \times t$$
$$L_c = \left(L_x^2 + L_z^2 \right)^{0.5} \tag{5.54}$$
$$t = \frac{v_{yik} + \left(v_{yik}^2 + 2g \times y_{ik} \right)^{0.5}}{g}$$

where $g = 3.986005 \times 10^{14}/[R_k(R_k + y_{ik})]$ and R_k is the earth radius of the impact point, which is set at the radius of earth. The formula for the calculation of y_{ik}, v_{yik} is as follows:

$$\varepsilon = \frac{L_x + L_z}{R_k}, \quad \beta = \arcsin\left(\frac{y}{R_k}\sin(\varepsilon)\right),$$

$$y_{ik} = \frac{R_k}{\sin(\varepsilon)}\sin(\varepsilon + \beta) - R_k, \quad V_{yik} = V_{yk}\cos(\varepsilon) \tag{5.55}$$

where y_{ik} and v_{yik} are the y-direction coordinate and v-direction speed of the starting point of the impact point in the ground coordinate system and the other coordinates are in the launching system. The coordinates of the impact point in the intermediate system is:

$$U_c = M_0 X_c + U_a = \begin{pmatrix} u_c & v_c & w_c \end{pmatrix}^T, \quad X_c = \begin{pmatrix} L_x & 0 & L_z \end{pmatrix} \tag{5.56}$$

We get the longitude and altitude of the ground λ_c, β_c from the impact point in the intermediate coordinate system.

Ellipse Model
When the height of local point is above 30 km, the ellipse model is used for calculations of impact point. Here are the calculations of parameters in the inertial coordinate system.

Forward ground coordinate system: the ground coordinate system in which the geodetic azimuth $\Phi = 90°$. We get the state parameter in the inertial system through the forward ground coordinate system. V_a in the launch coordinate is transferred forwardly:

$$V_k = M_k^T M_a V_a = \begin{pmatrix} v_{xk} & v_{yk} & v_{zk} \end{pmatrix}^T$$

$$|V_k| = \left[v_{xk}^2 + v_{yk}^2 + \left(v_{zk} + r_k\Omega\cos\Phi_k\right)^2\right]^{0.5}$$

$$\sin\theta_k = \frac{v_{yk}}{|V_k|}$$

$$\cos\theta_k = \frac{\left[v_{xk}^2 + \left(v_{zk} + r_k\Omega\cos\Phi_k\right)^2\right]^{0.5}}{|V_k|}$$

$$\sin\delta_k = \frac{v_{zk} + r_k\Omega\cos\Phi_k}{\left[v_{xk}^2 + \left(v_{zk} + r_k\Omega\cos\Phi_k\right)^2\right]^{0.5}} \tag{5.57}$$

$$\cos\delta_k = \left(1 - \sin\delta_k^2\right)^{0.5}$$

$$M_k^T = \begin{bmatrix} \cos\beta_k & -\cos\lambda_k\sin\beta_k & -\sin\lambda_k\sin\beta_k \\ -\sin\beta_k & \cos\lambda_k\cos\beta_k & \sin\lambda_k\cos\beta_k \\ 0 & -\sin\lambda_k & \cos\lambda_k \end{bmatrix}$$

$$\sin\lambda_k = \frac{W_k}{r_k\cos\Phi_k}, \quad \cos\lambda_k = \frac{v_k}{r_k\cos\Phi_k}$$

In the formula, θ_k is the angle between the velocity vector and the ground tangent plane below the substar, δ_k is the geodetic azimuth whose velocity vector is projected on the ground, and V_k is the velocity value under an inertial system. Furthermore, the latitude and longitude λ_c, β_c and range parameter L_c of impact point can be calculated by spherical trigonometry formulas. In order to ensure the accuracy of calculations, the parameters still need amendments.

5.2.3.2 The Calculations of Impact Points

The process of the calculation of impact points is shown in Figure 5.9. The input includes data measured by the measurement equipment, data processed by the computer for further calculations, and parameters of the missile in the launch coordinate system. Specifically, there are

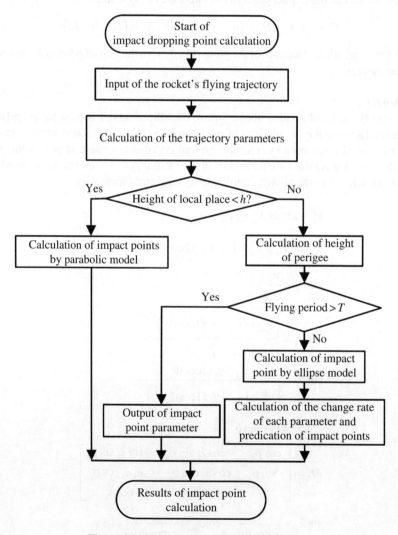

Figure 5.9 Impact point calculation flowchart.

position parameters X_k, Y_k, Z_k and speed parameters v_{xk}, v_{yk}, v_{zk}. And the output includes the velocity value, velocity angle, yaw angle, height of the local place, latitude and longitude of substar, latitude and longitude of the impact point, shooting range, and range of yaw of the carrier rocket in orbit at any instantaneous moment.

Through calculations of impact points, we get the rocket's flight status parameters. And by comparing these parameters with data of the security pipeline, we get to evaluate whether the parameters have exceeded the warning boundary and destruction boundary set by the security pipeline to get the necessary parameters for security estimation.

The results of impact point calculations will be used in real-time monitoring of impact points. In order to ensure the practicality of impact point calculations, the elliptic theory and correction are adopted for security estimation and aircraft fault conditions. The calculation adopts standard geophysical parameters published by the State.

5.2.4   Calculation Model of Shrapnel Distribution Range

When the rocket explodes, its shrapnel will have an enormous destructive power, leaving debris in various shapes and sizes, which fly at a different speed. The properties of the debris depend on the structure and explosion mode of the launch vehicle. The remote and external measurement data of the carrier rocket obtained by the tracking and telemetry system, combined with the terrain data provided by a geographic system, could be used for the estimation of rocket explosion model and determining its explosive power. According to the explosive power of the faulty rocket and the remote and external measurement data before an explosion, we could calculate the speed, acceleration, and other initial state of debris of different properties. Through force analysis of the debris, we get the differential equations of the flying debris, which, combined with the initial state and terrain data of debris, could be used in the calculation of differential equations to get the impact point of various explosive debris and its spreading area. The calculation and flowchart of the debris spreading area of the carrier rocket are shown in Figure 5.10. The application of the results of the calculations in GIS systems could visualize the debris spreading area and its impact over this area.

5.2.4.1   Determine the Explosive Status and Mode of the Carrier Launch

The Status of an Explosion
Based on the exterior ballistic measurement data and telemetry data received, the moment, location, and flight speed of the launched vehicle explosion are determined, and the quality of the remaining propellant and parameters that serve as the foundation for calculations of the rocket debris are also decided.

Explosive Mode of the Carrier Launch
Based on the exterior ballistic measurement data and telemetry data received in combination with the terrain data in geographic information systems, determine launch vehicle explosion mode, which could be divided into ground and air explosion according to the required explosive power of the vehicle.

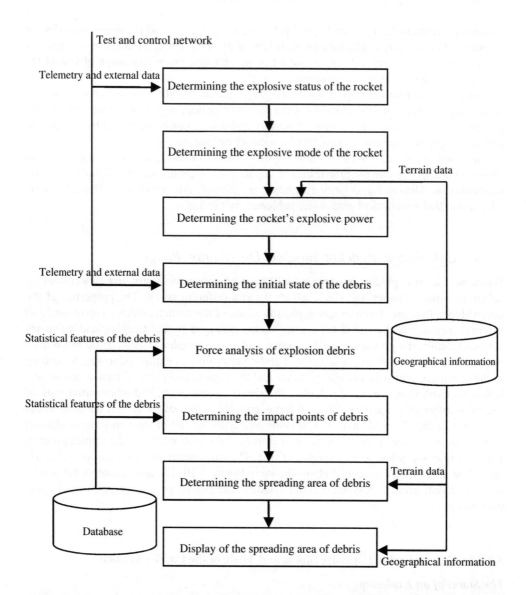

Figure 5.10   The calculation flowchart of explosion debris spreading area of the carrier launch.

5.2.4.2   Calculation of a Launch Vehicle Explosive Power

The formula for calculation of explosive power in the air:

$$M_{\text{TNT}} = M_{ln} \times 0.05 + M_{ll} \times 0.6 \tag{5.58}$$

The formula for calculation of explosive power on the ground:

$$M_{\text{TNT}} = M_{ln} \times 0.1 + M_{ll} \times 0.6 \tag{5.59}$$

In the formula, M_{TNT} is the TNT equivalent of the explosion, a component to be solved, M_{ln} is the normal remaining propellant of the carrier launch when it explodes, and M_{ll} is the remaining cryogenic propellant of the carrier launch when it explodes. After determining the explosive power, we get the parameters of the explosive blast, such as impulse i_s and lateral pressure p_s, which are used in the observation of how the impulses accelerate the debris.

5.2.4.3   Determining the Initial State of Explosion Debris

The explosion debris gets speed from two sources, one from the high pressure gas in the propellant storage box which accelerates the debris when the box explodes and the other from the explosive impulses.

The formula for calculation of the initial speed of debris under high pressure gas is as follows:

$$\begin{cases} u = \bar{U}ka_q \\ \lg \bar{U} = 1.2\lg \bar{P} + 0.91 \\ \bar{P} = \dfrac{(p - p_0)V_0}{M_c a_q^2} \\ k = 1.25\dfrac{m_p}{m_c} + 0.375 \end{cases} \tag{5.60}$$

In the formula, u is the speed of the debris under high pressure gas, a component to be solved, $\bar{U}$, $\bar{P}$, k are the intermediate variant, a_q is the sonic speed of gas produced by the explosion, p is the withstand voltage of the propellant storage box, p_0 is the air pressure of the explosion location, V_0 is the volume of the propellant storage box, m_c is the quality of the storage box, and m_p is the mass quality of the calculated debris. We get the speed u of debris under high pressure gas through calculating the known conditions and constant in the aforementioned formulas.

The formula for calculation of the speed acquired by debris under the explosive impulses is as follows:

$$v = \frac{p_0 i_s C_D A}{m_p p_s} \tag{5.61}$$

In the formula, v is the speed of debris under the impulses, a component to be solved, p_0 is the atmospheric pressure of the explosion location, i_s is the explosion-specific impulse, C_D is the resistant coefficient of debris, A is the area of thrust surface of debris, m_p is the mass quality of debris, and p_s is the lateral pressure of the explosive impulse. By calculating the known conditions in the formulas above, we get the speed v of debris under impulses.

After determining the speed u of debris under high pressure gas and speed v under impulses as well as the flying speed V of the launch vehicle at the moment of an explosion, it is possible to compose debris speed V_p projecting toward different directions, the formula of which is as follows:

$$\vec{V}_p = \vec{V} + \vec{u} + \vec{v} \tag{5.62}$$

5.2.4.4   Force Analysis of Explosive Debris and Impact Point Calculations

After analyzing the forces exerted upon the explosive debris in combination with the debris' speed direction on oxy, we get the following differential equations:

$$\begin{cases} \ddot{X} = -\dfrac{AC_{\mathrm{D}}\rho\left(\dot{X}^2+\dot{Y}^2\right)}{2m_p}\cos\alpha + \dfrac{AC_{\mathrm{L}}\rho\left(\dot{X}^2+\dot{Y}^2\right)}{2m_p}\sin\alpha \\[4mm] \ddot{Y} = -g - \dfrac{AC_{\mathrm{D}}\rho\left(\dot{X}^2+\dot{Y}^2\right)}{2m_p}\sin\alpha + \dfrac{AC_{\mathrm{L}}\rho\left(\dot{X}^2+\dot{Y}^2\right)}{2m_p}\cos\alpha \end{cases} \tag{5.63}$$

In the formula, X is the accelerated velocity of flying debris in the direction of x, X is the speed of debris when flying in the direction of x, Y is the accelerated velocity of flying debris in the direction of y, Y is the speed of debris when flying in the direction of y, A is the area of thrust surface of debris, C_{D} is the resistant coefficient of debris, C_{L} is the lift coefficient of debris, ρ is the density of debris, m_p is the mass quality of debris, and α is the angle of attack of the flying debris. As the initial position and initial velocity after the explosion debris are already known, the flight position of debris at various moments $(x_t, y_t, 0)$ could be calculated by iteration method and could be compared with the terrain data $(x_l, y_l, 0)$ in GIS. When $y_t = y_l$ in the iteration calculation, the debris has fallen on the ground, and the impact, point coordinate of the debris is $(x_l, y_l, 0)$.

5.2.4.5   Calculation and Display of the Spreading Area of the Explosion Debris

Determine the initial speed of various debris after an explosion and complete composition of velocities along different directions. Repeat the iteration calculations and record the impact points, and make statistical records of the impact points to determine the spreading area of the explosion debris and display it in GIS.

5.2.5   Spreading Calculation Model of Fuel Leakage Under by Wind Power

As to the determination of liquid-propelling launch vehicles falling and exploding area, scaling test is often adopted for this purpose both at home and aboard. The test results apply only to the rocket exploding on the launch pad, unsuitable for rockets that explode in the air. In this method, experimental data analysis is carried out, followed by the establishment of a mathematical model. And it also combines the wind speed during the flight of the spacecraft, weather and geospatial information such as prevailing wind direction in order to describe the wind fields over the complex terrain, calculations of poison gas concentrations over time, and simulation of poison gas spreading, as shown in Figure 5.11.

5.2.5.1   Initialization

Initialize the meteorological data (U, T, α) location (x, y, z) and time of flight (t_q), in which u stands for average wind speed, T is the absolute temperature of the air, and α is the prevailing wind direction.

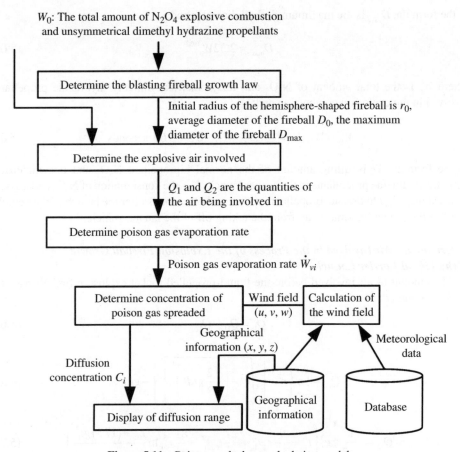

W_0: The total amount of N_2O_4 explosive combustion and unsymmetrical dimethyl hydrazine propellants

Determine the blasting fireball growth law

Initial radius of the hemisphere-shaped fireball is r_0, average diameter of the fireball D_0, the maximum diameter of the fireball D_{max}

Determine the explosive air involved

Q_1 and Q_2 are the quantities of the air being involved in

Determine poison gas evaporation rate

Poison gas evaporation rate $\dot{W}_{vi}$

Determine concentration of poison gas spreaded

Wind field (u, v, w) Calculation of the wind field

Geographical information (x, y, z)

Meteorological data

Diffusion concentration C_i

Display of diffusion range

Geographical information

Database

Figure 5.11   Poison gas leakage calculation model.

5.2.5.2   Determine Propellant Source Intensity

The determination of propellant source intensity is to determine the poison gas evaporation ratio, which is the input condition for the determination of poison gas spreading concentration.

Determine the Growth Rules of the Rocket Explosion Fireball's Liquid Propellant

The growth rule is primarily determined by the radius of the initial hemispherical fireball and its average diameter. The radius of the initial hemispherical fireball r_{00} is decided by the following formula:

$$r_{00} = 0.156 D_{max} \qquad (5.64)$$

The fireball average diameter D_0 is determined by a formula:

$$D_0 \approx 0.75 D_{max} \qquad (5.65)$$

In the formula, D_{max} is the maximum fireball diameter and:

$$D_{max} = 2.32 W_0^{0.32} \tag{5.66}$$

where W_0 is the total amount of N_2O_4 and unsymmetrical dimethyl hydrazine propellants involved in the explosion combustion and:

$$W_0 = W - U(t_q + 1.5) - W_e, \quad 0 \le t \le 60 \text{ seconds} \tag{5.67}$$

In the formula, W is filling amount of the normal bipropellant N_2O_4 and unsymmetrical dimethyl hydrazine propellants in the carrier launch, U is the consumption of N_2O_4 and unsymmetrical dimethyl hydrazine propellants per second 60 seconds after the launch of the vehicle, and t_q is the remaining time away from the taking off of the launch vehicle.

Determine the Air Involved in the Process of the Explosion Fireball Growth of the Liquid Carrier Launch

1. The amount of air involved before the liquid-propelled rocket explosion fireball rises off the ground Q_1:

$$Q_1 = Q_{11} - Q_{12} \tag{5.68}$$

$$Q_{11} = \int_0^{t_1} \pi \left\{ r^2 - \left[\sqrt{r^2 - \left(\int_0^t \frac{2}{3} gt \, dt \right)^2} \right]^2 \right\} \frac{2}{3} gt \, dt \tag{5.69}$$

$$Q_{12} = \frac{1}{243} \pi g^3 \left(t_1^6 - \frac{9}{2g} r_{00} t_1^4 + \frac{27}{g^2} r_{00}^2 t_1^2 - \frac{81}{g^3} r_{00}^3 \ln \frac{r_{00} + (1/3) g t_1^2}{r_{00}} \right) \tag{5.70}$$

2. In the formula, $t_{00} = 0.3329 W_0^{0.16}$, $t_{r00} = 0.1437 W_0^{0.16}$, and $t_1 = [(3D_0 - 6r_{00})/2g]^{0.5}$. In the formula, t_{00} is the time of the initial hemispheric fireball generates, t_{r00} is the period when the initial hemispheric fireball stops growing, t_1 is the period when the fireball grows from the hemisphere to a ball, g is the gravity acceleration with an approximate value of 9.8 m/second^2, π is the circularity ratio, t is the period after the fireball explodes, W_0 is the total amount of N_2O_4 and unsymmetrical dimethyl hydrazine propellants involved in the explosion combustion, D_0 is the average diameter of the fireball, and r_{00} is the radius of the initial hemispherical fireball.

3. The amount of air involved during the liquid-propelled rocket explosion fireball rises off the ground Q_{2t}:

$$Q_{2t} = \int_0^{t_2} \pi r^2 \left(V_{fbt} - \frac{1}{3} V_r \right) dt \tag{5.71}$$

In the formula, t_2 marks the period during which the fireball starts to rise until it reaches the maximum diameter; r and V_r are the radius and radial velocity of the fireball from its starting stage to rise until it reaches the maximum diameter, V_{fbt} is the rising speed of the fireball, t is the time of rising of the fireball, $0 \le t \le t_2$, and π is the circularity ratio.

Determine the Poison Gas Evaporation Ratio after the Liquid-Propelled Rocket Explodes

The evaporation rate could be determined by two ways, namely flash and steady evaporation rates based on whether the ambient temperature T (given) is above or below the boiling point when the propellant explodes.

1. Flash evaporation rates:

$$W_f = k_f \rho_{air} \left(Q_1 + Q_{2t} \right) \tag{5.72}$$

In the formula, k_f is determined by the ratio of oxygen in the air and the chemical reaction formula, 0.02028 for N_2O_4 and 0.009566 for unsymmetrical dimethyl hydrazine; Q_1 and Q_{2t} are decided by formula (5.68) and (5.71) respectively; and ρ_{air} is the density of the air.

2. Steady evaporation rate

$$\dot{W}_{vt} = \dot{W}_{v0} e^{-\frac{\dot{W}_{v0}}{W_f} t} \tag{5.73}$$

In the formula, $\dot{W}_{vt}$ is the steady evaporation rate of poison gas after the liquid-propelled carrier launch explodes, t is the period of evaporation, $\dot{W}_{v0}$ is the initial evaporation rate of the carrier launch's propellant, and

$$\dot{W}_{V0} = 0.03305 k_m m_f W_f G_f \tag{5.74}$$

$\dot{W}_f$ is the remaining mass quality of propellant in the fireball and calculated by formula (5.73).

5.2.5.3 Generation of Complex Terrain Meshes

In order to simulate the transmission or diffusion of the air flow that distributes vertically over the complex terrain, the terrain surface is taken as a mesh to accurately reflect the real effect of the topography, simulating hill fluctuations and the blocking, branching and streaming of the terrain, flow field status of the zone of wake of the terrain, that is, leeward slope whirl. The horizontal plane of the launch system coordinates (x, y) uses equal step length distribution while direction Z adopts the vertical coordinate transformation that reflects the topography, and the output is the body-fitted coordinates that fit the surface $x, y, \bar{Z}$. It can fully reflect the ups and downs of the terrain and may constitute a body-fitted curvilinear coordinate grid over complex terrain in order to determine the topographical mesh data of the wind field over the complex terrain, which is suitable for mountainous terrain. The mesh is decided by the formula below:

$$\bar{Z} = H \frac{Z - Z_g}{H - Z_g} \tag{5.75}$$

In the formula, H is the crest level at the point, $Z_g = Z_g(x, y)$ is the topographic relief height of the terrain, Z is the vertical coordinates of (x, y, z) in the Cartesian coordinate system, Z is the coordinate of crest level after the transformation; based on the formula (5.74), we get the range of Z: $\bar{Z} = [0, H]$.

After the segmentation of Z within $[0, H]$, $\bar{Z}$ is determined as:

$$Z = Z_g + \frac{\bar{Z}(H - Z_g)}{H} \tag{5.76}$$

5.2.5.4 Determining the Wind Field Over Complex Terrain

The numerical integration of equations is carried out after the explosion of liquid-propelled rocket propellant, such as forecasting future concentration. We know that the diffusion of concentrations is done under the propelling of wind. In the solution of diffusion equation, wind is a very important output and the foundation of the entire problem of poison gas diffusion. The wind field is determined by formulas (5.77) and (5.78).

$$\frac{\partial E}{\partial t} + u\frac{\partial E}{\partial x} + v\frac{\partial E}{\partial y} + w\frac{\partial E}{\partial z} = k_{mz}\left[\left(\frac{\partial u}{\partial z}\right)^2 + \left(\frac{\partial v}{\partial z}\right)^2\right] + \frac{\partial}{\partial x}\left(\frac{k_{mh}}{\sigma_E} \cdot \frac{\partial E}{\partial x}\right)$$

$$+ \frac{\partial}{\partial y}\left(\frac{k_{mh}}{\sigma_E} \cdot \frac{\partial E}{\partial y}\right) + \frac{\partial}{\partial z}\left(\frac{k_{mh}}{\sigma_E} \cdot \frac{\partial E}{\partial z}\right) - \varepsilon \tag{5.77}$$

$$\frac{\partial \varepsilon}{\partial t} + u\frac{\partial \varepsilon}{\partial x} + v\frac{\partial \varepsilon}{\partial y} + w\frac{\partial \varepsilon}{\partial z} = \frac{\partial}{\partial z}\left(\frac{k_{mz}}{\sigma\varepsilon} \cdot \frac{\partial \varepsilon}{\partial z}\right) + \frac{\partial}{\partial x}\left(\frac{k_{mh}}{\sigma\varepsilon} \cdot \frac{\partial \varepsilon}{\partial x}\right) + \frac{\partial}{\partial y}\left(\frac{k_{mh}}{\sigma\varepsilon} \cdot \frac{\partial \varepsilon}{\partial y}\right)$$

$$+ c_{1\varepsilon}\frac{\varepsilon^2}{E}k_{mz}\left[\left(\frac{\partial u}{\partial z}\right)^2 + \left(\frac{\partial v}{\partial z}\right)^2\right] - c_{2\varepsilon}\frac{\varepsilon^2}{E} \tag{5.78}$$

$$k_{mz} = c_u\frac{\varepsilon^2}{E}$$

In the formula, (u, v, w) are three wind speed vectors in the launch coordinate system to be solved, (x, y, z) are three components in the launch coordinate system transformed through formula (5.75) and (5.76) as given geographic information, E and ε are prevailing wind energy and momentum, respectively, determined by the direction of prevailing wind and wind speed, $c_{1\varepsilon}$ and $c_{2\varepsilon}$ are the second- and fourth-order dissipation coefficient respectively, and constant k_{mh} and k_{mz} are the wind field coefficient on the horizontal and earth elevation directions respectively. The simulated time compression correlation method is used to solve three wind vectors (u, v, w) in the launch coordinate system.

5.2.5.5 Determination of Poison Gas Diffusion and Concentration

In the process of toxic gas diffusion in the air after the liquid launch vehicle explodes, the diffusion speed and the three wind speed vectors (u, v, w) of the propellant after its explosion could be determined. Work out the diffusion equation; we get the diffusion concentration C_i of the toxic gas on the launch coordinate system (x, y, z) along with time vector t since the explosion of the propellant of the liquid launch vehicle, which serves as the concentration data of the area concerned. The concentration of toxic gas diffusion could be determined by Formula (5.77).

$$\frac{\partial c_i}{\partial t} + \frac{\partial}{\partial x}(uc_i) + \frac{\partial}{\partial y}(wc_i) + \frac{\partial}{\partial z}(wc_i) = D_i\left(\frac{\partial^2 c_i}{\partial x^2} + \frac{\partial^2 c_i}{\partial y^2} + \frac{\partial^2 c_i}{\partial z^2}\right) + R_i(c_i, T) + \dot{W}_{vi}(x, y, z, t)$$

(5.79)

where t represents the time since the explosion of the launch vehicle propellant; C_i is the diffusion concentration to be solved with parts per million volume concentration; D_i is the ith known molecular diffusion coefficient of gas composition; R_i is the chemical reaction rate of the ith ingredient, $\dot{W}_{vi}$ is the poison gas of ith component gas evaporation rate; T is the absolute temperature; (u, v, w) are the three known wind speed components in the coordinate system for launch.

5.3   Intelligent Decision of Space Flight and Safety Control

With the development of information technology, computers, and control technology, the ground safety control systems of space launching are gradually evolving into intelligent directions. Information, including trajectory, attitude, and impact point parameters, is in combination with space analysis of the impact point area in which the GIS system could achieve real-time spatial descriptions of security channel. The impact point forecast, flight parameters, and flight tracks are determined by the status of the launch carrier in the process of flight and they can provide real-time intelligent decision for the safety control of the launch vehicle.

5.3.1   Intelligent Decision Based on Spatial Information Processing

The Decision Support System (DSS), with the model base system as the main body, offers supplementary decisions through quantitative analysis. The nature of DSS is to combine various types of generalized models together to compose a problem-solving model system to offer information support for decision-making through processing the data in the database.

The knowledge reasoning technology in artificial intelligence (AI) and the basic functional modules of DSS are combined to form the Intelligent Decision Support System (IDSS). It not only gives full play to the characteristics of the expert system based on qualitative analysis of problems in the form of knowledge reasoning but also takes the DSS model as the core features solving quantitative problems, fully combining qualitative and quantitative analysis for considerable improvement of its problem-solving ability and capacity.

The IDSS, based on GIS, uses AI and GIS technologies in combination with other technologies, which include advanced software engineering, data warehouse, oriented object technology, super text and media, network and remote communications. Moreover, data mining and knowledge nuggets technologies are essential to strengthen space data and model analysis. The capacity on the basis of existing IDSS is making the decision behavior more scientific, visual, and intuitive. The integrated GIS module in IDSS architectures based on an intelligent support system of the spatial information is shown in Figure 5.12.

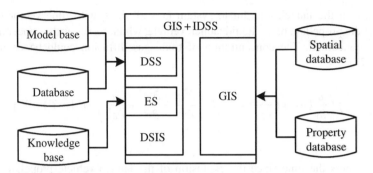

Figure 5.12   The integrated structure of GIS and IDSS.

From Figure 5.12, it is clear that the integration of IDSS and GIS plays a vital role in the establishment of the spatial intelligent decision support system (SIDSS). It features the following:

1. IDSS at the core: With DSS as the core, it extends its ability to space analysis such as supporting bidirectional search, adding graphics data management capabilities, and providing a variety of operations on the map. Apart from similar basic structure with general IDSS, it also has graphics query and space analysis functions in model management and data management as well as a space database management system to regulate related view manipulation and graphics data exchange with another system. At the same time, in order to enable the system to distinguish the processed object of each command (graphics, text, and model), system offers a human–computer interaction system, a module that makes the menu selection, utilization of data (text and graphic) by the model, various graphic manipulation, and query certain sequences of commands, which are then distributed to various functional subsystems and sent back to the caller.
2. SDSSs with GIS at the core: In a system of this type of integration, GIS not only provides an intuitive platform for data analysis and presentation with the DSS but also offers functions like input, organization, preprocessing, and postprocessing of spatial distributed parameters for the multimodel establishment in the DSS. Expert system in ISDSSs is the control center of the entire system, organization, and coordination of all parts of the system work, assigning various resources associated with user-defined issues, such as organizing information flows between environmental thematic databases and mathematical model of environmental, automatic control system operation sequence and models combined.
3. The spatial information system design model based on data analysis and functional decomposition: This design focuses on both spatial data analysis and systems feature modular decomposition, and the two are closely linked together by data flow concrete algorithm of the model, beneficial for the development of the object-oriented system interface, especially when integrated with the GIS system. The interaction between data analysis and functional decomposition as well as entity-relationship ER Figure of the system defined the data flow figure as "top to bottom" to offer practicality of systematic data and mathematical modeling. And it gradually clears the logical and physical structure of the spatial database as well as data classification, coding, and design of data-storage space of the system.

IDSS are often called expert system in a broad sense. Therefore, it also features the intelligent properties of ES in addition to its traditional decision support functions of DSS model. IDSS's main functions are as follows:

1. General reasoning: With the knowledge reasoning structures, the IDSS could simulate the thinking of policymakers to guide decision makers to choose the right model through questioning sessions, problem analysis, and rule-based reasoning based on experience and knowledge. However, when a model or algorithm fails to solve the problem, IDSS can assist policymakers to combine the existing models or help establish new models to solve the problem and carry out test control in model construction and operation.
2. Intelligent interfaces: It has the perfect human–machine interface (NLI or session interfaces) to guide decision-makers to determine the boundary conditions and the environment of a problem by way of a challenge response so as to effectively solve semi-structured and unstructured problems. The explanation mechanisms of IDSS could answer the decision makers' questions such as "What if," "Why," "When," and "How" to provide decision makers with means of problem tracking and solving process and increase the credibility of the results.
3. Self-study: IDSS has self-learning capability, which can automatically analyze the decisions of policymakers in the decision-making process, extracting policy-making experience knowledge and heuristic rules, perfecting case library, and eventually supplementing the decision-making process automatically.

IDSS, with information technology as the means, uses theories and methods of decision science and related disciplines and combines quantitative analysis and qualitative analysis to solve structured, semi-structured, and unstructured problems. Therefore, the IDSS are widely used in the area of decision-making.

Based on the aforementioned analysis, we can see that in the area of decision support, GIS and IDSS have its own advantages and disadvantages. If GIS combines spatial data processing and representation ability with IDSS's solution of semi-structured and unstructured problems, the two could perfectly complement each other. The integration of IDSS and GIS not only provides spatial data processing capabilities and more vivid visualized platform for IDSS but also enables GIS to transform from description and representation of the objective world to a new stage that directly take part in and transform the objective world.

5.3.2 Structure of SIDSS of Space Flight Safety Control

Through the integration of information technology and GIS technology, the structure uses radar, optical, and other trajectory measurement systems to acquire real-time trajectory data as well as e telemetry subsystem to get real-time flight status and trajectory data of the launch vehicle. Through data processing, the results are combined with GIS for analysis and display. When a failure occurs, the space or properties' information are extracted from the GIS to serve for model analysis and calculations of related geographic information data to estimate losses and analyze various aspects of the situation effectively. Based on the location and severity of the accident, the system offers corresponding faults treatment plan through decision-making process and provides needed information and decision support to visualize the emergency responding decisions with adequate information and intelligence. The structure is shown in Figure 5.13.

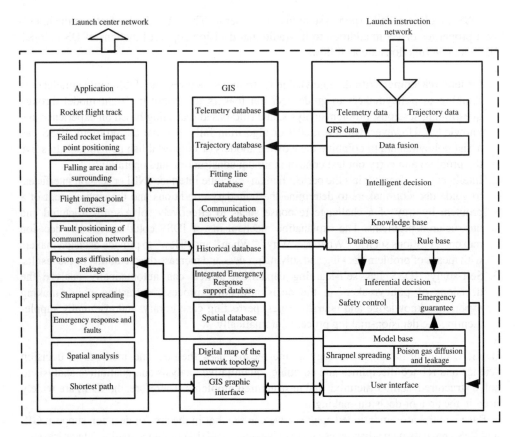

Figure 5.13   Flowchart of space intelligent decision support system of space flight safety control.

Digital maps are an important carrier of GIS, providing necessary spatial and property information for the system to complete its functions.

The system converts the data to a graphics path by map spatial data, which describes the real-time vehicle trajectory data under transient state of the launch vehicle and overlays it with the established launching site and the digital map of the entire navigating zone to dynamically reflect the rocket's flight in real time. The establishment of GIS is an important step for space analysis on safety control, providing geographic entities and properties information required to complete safety control features. Geo-spatial database and attribute database should be established in accordance with the guidelines that facilitate safety control and emergency support to ensure that the functioning system gets property information of related geographical entities in the space environment and achieves effective decision and emergency information support.

The spatial data processing capability of dynamic GIS is to offer decision makers information such as speed, position, height, subastral location, impact point location, and distance from the protected target and launch vehicle status. The GIS gives vivid descriptions of related ballistic parameters of the launch vehicle and working conditions to facilitate the launch vehicle to make the right decision in terms of its flight status.

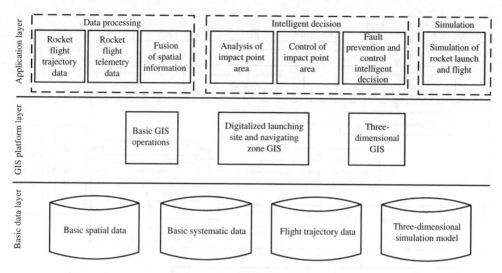

Figure 5.14   Systematic structure.

In the impact point area part of GIS, the database stores spatial and property information of cities and facilities in different time periods to achieve multidimensional GIS, space geometric operations such as GIS overlay, and dynamically display the target data information in GIS, offering references for the detailed design of the impact point area and selection. The systematic structure is shown in Figure 5.14.

5.4   Intelligent Decision-Making Method for Space Launch and Flight Safety Control

During the flight and launch process of satellite, the method targets safety control and emergency support for space launch and takes advantage of GIS spatial information descriptions to describe the surroundings of the faulty point and the faulty area through simulation and space information processing, simulate the launch and flight of satellite, and analyze the site field, providing intelligent decision support and emergency response information.

5.4.1   Space Launch and Flight Safety Control Decision

5.4.1.1   Flowchart Analysis of Space Launch and Flight Safety Control

The space flight safety control decision system is a computerized IDSS that integrates rocket flight status monitoring, trajectory optimization, data processing, and security incident treatment. Knowledge and experience of space flight safety control is combined with the ballistic data processing and error analysis to constitute a knowledge base and model library. Then the estimation mechanism of safety control, together with artificial intelligent techniques, constitutes the reasoning mechanism of the system. According to real-time flight status parameters of the launch vehicle, we get relevant knowledge and models associated

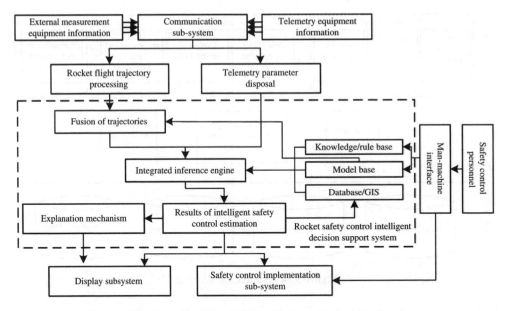

Figure 5.15   Space launch and flight safety control decision flowchart.

with trajectory data and calling as well as safety-related information to choose the right call control scheme through an integrated inference engine.

Based on the ideas of Figure 5.15, it offers the system processing flowchart. In the IDSS for safety control of the launch vehicle, the system describes and stores a variety of quantitative computation models. Moreover, empirical knowledge of safety control strategies is for the implementation of launch safety control and management in the form of databases (spatial database, property database, flight database) and model libraries.

From Figure 5.15, it is clear that the process of space launch and flight safety control decision system could be divided into three steps:

1. In the processing of launch vehicle's measurement information, the position and speed information of the rocket flights measured under various coordinate systems is transferred into flight status information under uniformed coordinate system and time sampling point. The coded telemetry data is processed into a telemetry parameter of each subsystem of rocket flight, marking the completion of security decision's parameters and data prepro-cessing required in a real-time flight.
2. Through fusion of multisourced data, we get trajectories of higher precision and reliability. The telemetry parameters are utilized in the process for further analysis of current rocket flight status. Based on the needs of security estimation, experience and knowledge of space launch and flight safety control (reasoning rules) and various calculation models (such as equipment error models and credibility distribution model), the system produces real-time flight security results and correspondent executive decisions through quantitative model calculations and qualitative knowledge reasoning.
3. The safety control personnel select the final implementation program through human–machine interfaces, smart decision-making support system, and their own decisions. According to

the established safety control policies and safety controls programs, the security personnel command the control equipment on the ground to make various safety instructions, or even order the faulty rocket to be destroyed.

In the sections above, the decision system plays the core of the entire launch and flight safety monitoring management.

5.4.1.2 Design of Functional Modules

From the figure of space launch and flight safety control decision flowchart shown in Figure 5.15, we see that it consists mainly of 11 modules, including information accessing, trajectory processing, telemetry data processing, knowledge/rule base, model library, database/GIS, integrated inference engine, explanation mechanism, information display, human–computer interface, and safety control implementation.

1. Information accessing: It accesses real-time flight external measurement information and telemetry data through exchange between the communication subsystem and monitoring information network.
2. Trajectory processing: According to the actual measurement system and trajectory data in combination with the correspondent optimized methods and models, the system transforms the multiple redundant external measurement equipment information into multiple external measurement trajectories and complete space and time alignment, in which the telemetry trajectories under inertial system and GIS trajectory under a geocentric system are aligned with the external measurement trajectory in terms of both time and space.
3. Telemetry data processing: Process the telemetry data concerned in a real-time manner according to intelligent security support decisions.
4. Knowledge/rule base: It is the basis of the decision system, which describes the knowledge of flight safety control of the launch vehicle and expertise, including security property parameters, facts, and safety control rules.
5. Model library: It includes models for trajectory data analysis and optimization and data fusion processing, GIS calculation and analysis model as well as models for analysis and calculation of launch vehicle explosion and poison gas diffusion.
6. Database/GIS: It is responsible for organizing and managing theoretical trajectories, security channels, and geographical information, and stores correspondent real-time safety estimation results and related real-time data, including impact points, speed, position, and pressure. The value of these parameters is different at each decision-making moment, making it a dynamic database.
7. Integrated inference engine: It consists of qualitative knowledge reasoning model and model calculation. Knowledge reasoning technique is based on the decision-making network, in which the engine produces various estimation results and safety control schemes through unidirectional or bidirectional searches based on the security properties.
8. Explanation mechanism: It offers the explanation of various safety estimation results for the security personnel, including data and time, a conclusion of reasoning, and the basic facts involved in reasoning.
9. Display of information: The module integrates the detected safety situation with a theoretical flight trajectory, spatial-temporal coordinates, safety channel, and geographic

information and displays the safety situation of the rocket at each moment and the real-time estimation results so that the safety personnel could understand the overall safety situation in an easy and intuitive way and make appropriate safety control decisions according to the actual situation.

10. Human–computer interface: The module existed in the whole process of analysis, decision-making, and control, through which the operators utilize the knowledge base and model library or search for relevant data and decision-making information. The system fully uses various data, model, and knowledge in combination with data processing, model calculation, and knowledge reasoning to select a safety control decision plan of high reliability so as to assist the safety personnel with the final decision.

11. Safety implementation: The module delivers the result of safety estimation to the correspondent safety control device under manual interference.

5.4.2 Minimum Loss-Based Space Launch and Flight Safety Control Real-Time Decision

The safety control decision-making system is a real-time security decision-making system with feedbacks based on spatial information and loss estimation. The decision-making control is significant because it aims at seeking timing and location of minimum loss by processing the spatial information of the accident location on the computer, as shown in Figure 5.16.

The real-time decision, in a broad sense, refers to computer processing and control, which detects the instantaneous value of the controlled parameter and decides further control based on input and analysis. At the site of an explosion of the faulty rocket, the system could determine the shrapnel dispersion and gas diffusion in the first place through spatial and geographical distributions and reflect the dynamic characteristics of the development of the accident to estimate the risks and offer the situation and trend of the accident to the decision maker through GIS in a timely manner. The information includes the accident location, impacts caused by the accident, and the shortest paths to the accident location.

According to the launch vehicle flight's track of subsatellite point, combined with GIS-spatial information, a comprehensive prediction of the impact point's location, hazard, and other factors affecting the rocket's safety control could be offered, upon which we get the time-series for carrier rocket flight control.

Flight moment k: $1, 2, \ldots, N$.
Impact point D: $D_1, D_2, \ldots, D_N \in D_i(x_i, y_i)$, $i = 1, 2, \ldots, N$;
Shrapnel distribution R: $R_{1,h1}, R_{2,h2}, \ldots, R_{N,hN} \in R_{i,hi}(x_i, y_i)$, $i = 1, 2, \ldots, N$;
Gad diffusion C: $C(h < h_0)$.

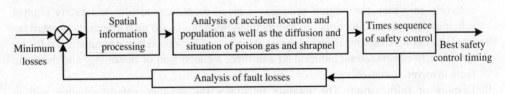

Figure 5.16   Safety control decision.

Population distribution P: $P = \rho \times R_{i,hi}(x_i, y_i)$, where ρ is the population density of the impact point coordinated. The decision value is min (P).

Important target σ: $\sigma(x_i, y_i) \cap R_{i,hi}(x_i, y_i) = Q$. $\sigma = \{$large infrastructure, river, lake$\}$, $\sigma(x_i, y_i)$ represents a protective target at (x_i, y_i). The decision value $Q = $ null or $Q = \min(Q)$.

5.4.3   Knowledge Representation of Decision

The various states of rocket flight is essentially the fusion of states of characteristic parameters during the flight, which uniquely determines the rocket's flight status. The following formula represents the relationship between flight characteristic parameters and the state of flight:

$$\Phi_k = \left\{ S_1, S_2, S_3 \mid t = t_k \right\} = \left\{ S_1(k), S_2(k), S_3(k) \right\}, \quad s_j(k) \in \left[s_{j1}(k), \ldots, s_{jn}(k) \right] \qquad (5.80)$$

In the formula, Φ_k is the flight status of the launch vehicle at moment t_k and $S_j(k)$ $(j = 1,2,3)$ is the characteristic parameter of flight status at moment t_k. The two are the three-dimensional location coordinate and the flight position parameter respectively. $S_{jl}(k)$ $(l = 1,\ldots,n)$ is a value of $S_j(k)$. At moment t_k, $S_j(k)$, fusion of the characteristic parameters of flight status, describes the status Φ_k at this moment.

The coordinates and position parameters of the rocket flight are decomposed into feature state components under various parameters. Classify formula (5.78) of real-time feature state components of flight status of the launch vehicle according to priority and general class and store the data in the knowledge library.

The feature state components of flight status of the launch vehicle could be described as a triple:

$$T = \left\langle \Phi, Q, \varphi \right\rangle \qquad (5.81)$$

In the formula, Φ is a finite set $\Phi = \{s_1, s_2, s_3\}$ of real-time feature state components of flight status; Q is the normal state domain set $Q = \{q_{s1}, q_{s2}, q_{s3}\}$, $q_{si} = [q_{si,1}, q_{si,2}, \ldots, q_{si,n}]$; and φ is the flight safety function domain set $\varphi_s = \{\varphi_s(q_{s1}), \varphi_s(q_{s2}), \varphi_s(q_{s3})\}$ under status Q during the flight of the launch vehicle.

We call the formula (5.79) the rule-based feature state components. The rule base is defined as basic information set, which is concluded as True. When the real-time flight status Φ_k is within the normal range of φ under status Q, the launch vehicle is considered to be working under normal conditions.

The structure of the rule-based feature state components is represented as:

Flight feature Φ_k of the launch vehicle	Normal flight status Q	Normal range of values φ of the launch vehicle at the present moment

In the space safety and emergency decisions, we put forward the domain knowledge representation model based on "goal–rule based feature state components." The domain knowledge can be divided into goal, a set of basic information concluded as "True" (rule based) and facts, which include corrections of different factual characteristics that form the domain rules.

5.4.4   Space Flight Safety Judgment Rules

By integrating the external measurement and telemetry parameter of the launch vehicle to judge whether the vehicle has crossed the warning line and the destruction line, we could decide if the rocket flight trajectory parameters is within the normal range, acquiring the design principles and judge rules of flight safety for the launch vehicle.

5.4.4.1   Design Principles of Flight Safety Control of the Launch Vehicle Assumptions

1. When the impact point of the external measurement trajectory parameters crosses the safety channel, the parameter of the telemetry position control system serves as the overall reference for judgment;
2. When the shooting range of the external measurement trajectory parameters crosses the safety channel, the telemetry position control system and the telemetry parameter of the dynamic system serve as the overall reference for judgment;
3. When the space location of the external measurement trajectory parameters crosses the safety channel, the telemetry parameter of the telemetry position control system serves as the overall reference for judgment;
4. When the flight velocity of the external measurement trajectory parameters crosses the safety channel, the telemetry parameter of the dynamic system serves as the overall reference for judgment.

5.4.4.2   Launch Vehicle Flight Safety Judgment Rules

According to the knowledge rule design's principles, when abnormal conditions arise in the trajectory parameters of the launch vehicle and cross the warning line, the rules should clarify the cause of warning, that is, the specific parameters that cause the abnormal conditions.. Now take the external measurement impact point parameter, shooting rage parameter, and velocity that cross the boundary as the example to illustrate the flight safety knowledge rule of the launch vehicle.

The real-time flight status Φ_k could be represented by $\Phi_k = \{S_1, S_2, S_3\}$.

Judgment of Cross-Boundary for Impact Point Parameter
The impact point boundary crossing is defined as $S_1(k)$:

$$S_1(k) = \{S_{11}(k), S_{12}(k), \ldots, S_{17}(k)\}$$
$$S_{11}(k) = W_{\beta c}, \ S_{12}(k) = \beta_c, \ S_{13}(k) = T_{\beta c}, \ S_{14}(k) = J_{\beta c}, \ S_{15}(k) = \alpha, \ S_{16}(k) = \beta, \ S_{17}(k) = \gamma$$

In the formula, $W_{\beta c}$ is the cross-boundary warning of external measurement impact point, β_c is the external impact point, $T_{\beta c}$ is the period during which the external impact point crosses the warning line continuously, $J_{\beta c}$ is the cross-boundary warning line of external measurement impact point, α is the pitching angle of the launch vehicle, β is the yaw angle of the vehicle, and γ is the roll-off angel of the vehicle.

The change in the impact point is related to the position parameter of the control system. By combining analysis of the effects of the telemetry parameter on the launch vehicle's flight results, the system makes an integrated estimation of the primary and secondary control system parameters of the telemetry measurement to further determine the reliability of the security results. When at least two control system parameters are found abnormal in the same period, the cross-boundary status of the impact point is determined as "True."

Judgment of Cross-Boundary for Impact Point Parameter

The shooting range parameter boundary-crossing is defined as $S_2(k)$:

$$S_2(k) = \{S_{21}(k), S_{22}(k), \dots, S_{27}(k)\};$$

$$S_{21}(k) = W_{Lc}, \; S_{22}(k) = L_c, \; S_{23}(k) = T_{Lc}, \; S_{24}(k) = J_{Lc}, \; S_{25}(k) = P, \; S_{26}(k) = \theta_1, \; S_{27}(k) = \theta_2$$

In the formula, W_{Lc} is the cross-boundary warning for external measurement and shooting range parameter, L_c is the shooting range, T_{Lc} is the period in which the external measurement shooting rage crosses the boundary continuously, J_{Lc} is the warning line for external measurement shooting rage, P is the combustor pressure of the engine of the vehicle, θ_1 is the positive pitch angle of the vehicle's velocity, and θ_2 is the negative pitch angle of the vehicle's velocity.

According to the analysis, the shooting range of the carrier rockets is related with the instantaneous velocity and attitude of the vehicle. Therefore, the judgment consists mainly of dynamical systems, control systems, and rocket pitching changes as well as the power system parameter and the change of carrier rocket pitch. Small thrust force generates short shooting range and fierce thrust force generates long shooting range; when the flight pitching angle is close to 45°, the shooting range is long and vice versa. In the judgment of the system, when the shooting range parameter is beyond the warning line, it first calculates the gap between actual and theoretical shooting ranges while the flight pitching angle changes along or against 45° and decides the correspondent telemetry power system parameter is conducive to increase or reduce the thrust force. When either of the telemetry pitching parameters or the dynamic power parameters share the same deviation direction within the shooting range of the vehicle, the status of shooting range cross-boundary is deemed as true, or vice versa.

Judgment of Cross-Boundary for Velocity Parameter

The velocity parameter boundary crossing is defined as $S_3(k)$:

$$S_3(k) = \{S_{31}(k), S_{32}(k), \dots, S_{37}(k)\}$$

$$S_{31}(k) = W_{Vk}, \; S_{32}(k) = V_k, \; S_{33}(k) = T_{Vk}, \; S_{34}(k) = J_{Vk}, \; S_{35}(k) = P_1, \; S_{36}(k) = P_{yx}, \; S_{37}(k) = P_{rx}$$

In the formula, W_{Vk} is the cross-boundary warning for the external measurement velocity parameter, V_k is the flight speed, T_{Vk} is the period during which the external measurement speed crosses the boundary continuously, J_{Vk} is the warning line for cross-boundary external measurement speed, P_1 is the pressure within the thrust chamber of the launch vehicle, P_{yx} is the pressure of the reserve tank of oxidizing agent, and P_{rx} is the pressure of the reserve tank of an incendiary tank.

Changes of the speed parameters are mainly concerned with the changes of power system parameters. Furthermore, when the thrust force of the launch vehicle is small, the speed is low and big thrust force would induce higher speed; when the speed parameter is beyond the warning boundary, the system first calculates the gap between the actual and theoretical speed and decides if the correspondent telemetry power system parameter is conducive to increase or reduce the thrust force. When the results of the power system parameter match with the rocket speed deviation direction, the status of cross-boundary of the velocity parameter is determined as "True," or vice versa.

5.4.5 Space Flight Safety Control Inferential Decision

The inference of space flight safety involves two steps: first, determine whether the safety parameters are in abnormal conditions (cross-boundary); second, determine various safety control instructions based on the status of various parameters.

When determining whether a safety parameter has crossed the boundary, it first calculates the parameters according to the general principles of safety inference, and if the results prove the parameter has crossed the boundary, then according to the reverse inference mechanism, the parameter is probably beyond the boundary. And with these results as assumptions, the system carries out reverse reasoning to look for facts that support this hypothesis. The principles for safety inference are stated in Section 5.4.3.

As to the reasoning of the safety control instructions, we set up three data tables: safety inference table, rule meta table, and rule table. Apart from the overall logic relation between the three tables as shown in E-R Table, each table is not completely independent from each other. Figure 5.17 illustrates the relations between each table while reflecting the logic structure of safety inference knowledge.

5.4.5.1 The Mechanism of Reasoning Decision Network

According to the representation forms of the safety inference knowledge, we introduced the concept of a tree to set up correspondent safety inference decisions network. Here, we take the "general alarm" model as an example to build the network. Every piece of knowledge is defined as follows: There are more than two knowledge prerequisites; the combination between the prerequisites is represented only by "logic and" and "logic or"; and there is only one conclusion. Therefore, in the decision-making process of the safety controls, the problem-solving process can be represented by an "and/or" tree. Refer to the knowledge representation of safety controls decision described in this chapter as well as the rocket flight safety rules in which we form the "general alarm" reasoning and decision-making network as shown in Figure 5.18.

5.4.5.2 Inference Controls Decisions

The reasoning process is a process of thinking, namely a process of solving problems. The quality depends not only on the problem-solving methods but also on the strategy for solving problems, that is, the inference control strategy. The inference control strategies mainly include reasoning, conflict resolution strategies, and search tactics.

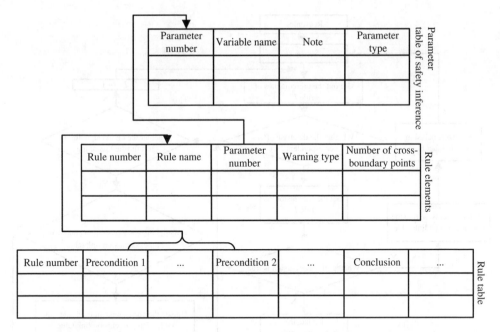

Figure 5.17   Relation schema of each data table in safety inference knowledge library.

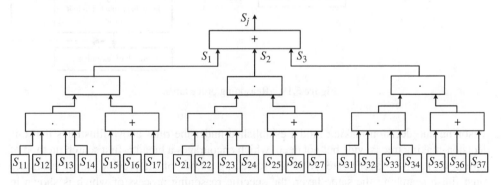

Figure 5.18   "General alarm" decision-making network.

The reasoning system keeps matching the currently known facts with knowledge in the knowledge library. It may happen that there is more than one knowledge prerequisite that meets the requirements, that is, all the prerequisites match with the facts, leading to conflicts between these prerequisites. How to choose the rule for execution becomes the main task of the conflict-resolution strategy. In the safety decision-making process, every rule plays a very important role in the safety inference and cannot be ignored. The rules whose preconditions match with the facts should be activated and be resolved by the conflict-resolution method.

Search is a fundamental problem in AI, an integral part of the reasoning. It is directly related to the performance and efficiency of the intelligent system. The so-called search strategy refers to seek the best path of reasoning method's reasoning under certain circumstances, which are divided into blind search and heuristic search. The "external measurement alarm"

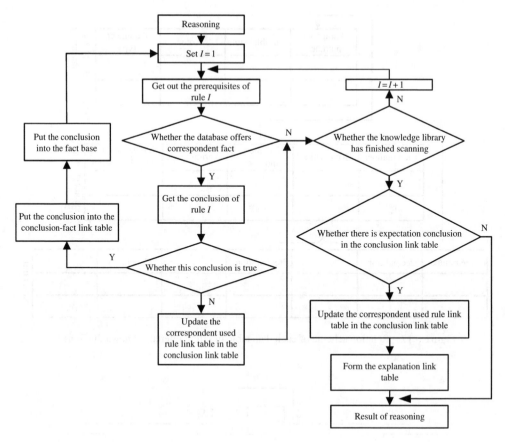

Figure 5.19   Reasoning procedures.

reasoning and decision-making model established under the rules above illustrates that this decision-making tree is wide but not deep, which encourages a breadth-first search method of bottom-up layer-by-layer strategy; before searching a node of the previous layer, all nodes must finish searching the same layer, the specific reasoning process of which is shown in Figure 5.19.

5.4.5.3   Inference Explanation Mechanism

There are huge differences between DSS and traditional logical software execution behaviors. As programs are executed in order in the traditional software, the programmers could eliminate the hidden errors by observing the behavior of the system. In the expert systems, as the knowledge library and reasoning machine are separated, the order of an execution system of this inferential knowledge is unpredictable because the activation of a rule is determined by both the prerequisites and inference engine. In order to get a clear understanding of the space launch safety controls mechanism based on knowledge inference, an explanation of the reasoning process is provided here.

The interpretation mechanism provided by the system is as follows: In the implementation of a safety decision process, the system outputs the inference results as well as produces and stores correspondent reasoning explanation. In the system, the reasoning explanation includes date and time of inference, reasoning conclusion, and the basic facts used by reasoning and other information, which are stored as in the form of database. When a particular conclusion is acquired through reasoning, if the user wants to understand how the conclusion is obtained, this mechanism could offer an inference explanation to the user about how the conclusion comes to improve the understanding of users; it also increases the transparency of the system, making it easier for users to accept the inference results. If the users are experts in the field, the display of the reasoning process could help him understand the work and rationality of knowledge library and is beneficial to system maintenance.

5.5   Safety Emergency Response Decision of Space Launch and Flight

5.5.1   Structure of Intelligent Emergency Response Decision System

The space flight safety and emergency decision combined the IDSS and GIS technologies to establish an ISDSS based on plan library. The overall structure is shown in Figure 5.20, in which the system is composed of a database system, model base system, knowledge base system, GIS system, library system, plan subsystem, integrated reasoning system, human–computer interface, and decision-making subsystem.

Geographic information subsystem: To play basic functions, including spatial data collection, storage, management, analysis, simulation, and illustration. GIS visualizes the spatial data of the spatial database as the digital map which is displayed on the screen. In the GIS-related decision-making processes, the structure serves as the platform supported by spatial database and database properties to carry out analysis and simulation of space-related issues and offer more intuitive and vivid analysis, management, and decision-making tools for the entire system.

1. Database subsystem: The database stores the data associated with the emergency decision, such as analyzing model parameters and the intermediate results of decision-making calculation.
2. Model base subsystem: The model base stocks a variety of analysis models, including shrapnel dispersion model, poison gas dissipation model, and case matching model. These models are managed and maintained by the model base management system.
3. Knowledge base subsystem: The IDSS is intelligent because it has a decent knowledge base and reasoning engine. The knowledge base mainly stores knowledge related to emergency-responding decision-making, including preplan knowledge, preplanning evaluation model, evaluation criteria, and modification of the preplan.
4. Preplanning subsystem: It consists of preplan library, case library, and preplan library management system. The preplan library stores the preplanned emergency plans and the preplan collection forms the preplan library. Case library stores successful cases in the past. From the perspective of problem solving, preplans and case study should contain the overall descriptions of the problem as well as a description of the problem-solving approach, sometimes also should provide descriptions of the results. The preplan library management system is responsible for the maintenance and management of the preplans.

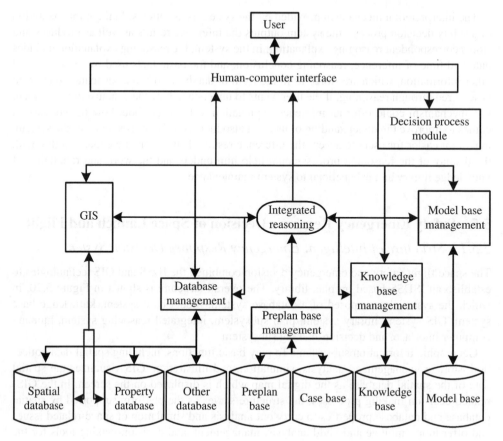

Figure 5.20   System structure.

5. Integrated reasoning subsystem: Implements the search for case study and selection of pre-plan decision based on rule reasoning with case reasoning at the core. The case library, which stores the proven decision methods, first matches the input problem with the proven case, and if the two match, the system offers an immediate resolution. However, as the accident, in reality, could be very different, it is very difficult to find a perfectly matching case. In case of failure of a direct case match, the system could adapt to be new of emergency needs by properly correcting the preplan by rule reasoning based on suitable preplan in the library through case reasoning.

6. Decision reasoning subsystem: Assist the human–computer interface in objective analysis and problem decomposition of a given problem, from the problem solving chain, check the legitimacy of decision-making tasks and issues, choose the right reasoning approach, and manage the coordination between various modules.

7. Human–computer interface: The human–computer interface is the interface between the whole system and the user. It provides users with friendly graphics interface and menu drive, understand problem's input by the user in various forms (main in the form of menu selection), and convert the problem into the form that the system can understand; in the running and reasoning process of the whole system, it allows the decision-maker to offer

direct suggestions and accept the subjective opinions and experience of the decision-maker: to display the results of system running in the form that the decision-maker is familiar with (such as table, and curve, statistics figure, and text description) and explain the results. The human–computer interface includes user interface and internal and external interfaces.

5.5.2   System Functions

According to the system analysis, the system requires the implementation of the following functions.

1. Information management: Input, store, process, and manage the spatial property data of buildings, equipment, and facilities of the launching district, spatial property data of the navigating zone, as well as the flight data in the process of space launch and flight. It provides efficient, accurate query and search means, offering information guarantees for emergency responding personnel. Through integration and combination of telemetry measurement, GPS, and modern mapping technology and information systems, it updates the geographic information in a real-time manner and ensures the temporality of information.
2. Support decision: Based on the state data and the geographical property data at the site of malfunction of the launch vehicle, the system assists the emergency personnel in developing emergency response plan and providing timely and effective decisions for managers at all levels to improve the space launch emergency support technology.
3. Visualized representation of information and decision–making: The visualization platform of GIS provides decision makers with flight information and plans of accident response through images, text, tables, and other techniques. The vivid expression improves the decision-making ability of the decision maker.

Based on the above analysis, the entire system can be divided into four modules: GIS modules, data-management module, emergency response module, and human–computer interactive module. The modular structure of the system is shown in Figure 5.21.

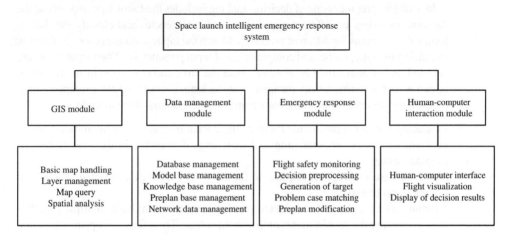

Figure 5.21   Functional modules of the system.

1. GIS module: The system provides the visualized emergency response platform whose features include basic map operation, layer management, map query, and spatial analysis.
2. Data-management module: Manage and process data of the system, including five libraries (database, the model library, knowledge library, preplan library, and case library), and send and receive network data from the server and client side. The server-side receives rocket flight data in real time from the control network, then processes the real-time data and sends the data after repackaging the data as the new data frame.
3. Emergency response module: In case of launch failure, the system assists the emergency personnel in making accident response decisions in accordance with population distribution, protection facility layout, as well as information of rivers and roads to provide decision support for the emergency response. Emergency module includes the following features:

   1. Real-time monitoring of flight safety: By monitoring the real-time rocket safety controlling parameters, the launch vehicle's safety is being monitored. In case of any fault, the system would make clear the fault type in order to activate preplans in a timely manner and offer emergency treatment and control information.
   2. Preprocessing of decision: There are three types of decision data needed by the system:

      1. Data A: Data that directly received by the system in case of fault, such as flight data, telemetry parameter, safety controls parameter, explosion type of the launch vehicles, real-time prevailing weather conditions, wind power, wind direction, etc.
      2. Data B: Manually input data, such as faults of the communication system, power system, filling system, etc.
      3. Data C: Data that requires reasoning and analysis of data source A and data source B, such as the effect of shrapnel distribution, poison gas diffusion, geographical property data within the range of the accident, etc.

      Though data A and B are the most direct information sources of the system, these data are very limited and far from enough to meet the needs of decision. Based on a variety of analysis model and GIS platform, the system gets data C that contains more in-depth information, such as spatial property data, the range of shrapnel explosion, scope of poison gas diffusion, and influence of faults on the ground, which is where the advantages of SIDSS lies.

      In addition, preprocessing of decision-making includes the following: decompose the decision according to different requirements of emergencies and classify the data to form solving chains for different problems. Match the input parameters for verification, examining the type, scope, and completeness of input parameters. When input errors are detected, the system would be alarmed about the error and correction instructions will be carried out; when necessary parameters are lacking, it would suggest prompt completion and coordinate works within the organizational systems of the module.

   3. Producing the target problems: Produce the decision-targeted problem based on the problem and carry out searching and matching of preplans in the case library or preplan library.
   4. Match of case or preplan: Explain to the user their queries and finish searches of preplans and actual cases.
   5. Estimation of preplans: The selected preplan sources might have multiple possibilities in meeting the requirements of threshold values. Therefore, a preplan-estimation approach is required in order to find the best preplan.

6. Rewriting and study of preplan: Modify the selected source cases to make it meet the problem-solving requirements of the current problem and get the final preplan. If the plan proves an absence in the case library, but is of great value, then the plan will be stored in the case library as a proven case.

4. Human–computer interface: As an important part of SIDSS, it can provide decision makers with a convenient, friendly, and interactive environment and intelligent human–computer interface. It provides flight data, flight path, and safety controls information during rocket launch while adopting heuristic knowledge and expertise in the process of case selection and production of a new plan. This section generally has the following functions:

1. Human–computer interface: Be responsible for interaction with the user, including providing with the decision makers the authority for system operation and model selection; guide the users with preplan information in an inspiring manner.
2. Visualized flight process of the launch vehicle: The system draws and overlay the rocket's flight path on GIS according to the flight data of the launch vehicle while visualizing state parameters of the launch vehicle, such as real-time ballistic data, real-time telemetry data, safety control data, time data, and the safety information of protected cities in a variety of ways in order to provide decision makers with vivid and intuitive decision-making information.
3. Display of the result of decision-making: Display the result in the form of table, text, image, and map.

5.5.3 Space Flight Safety Intelligent Decision-Making and Its Application

In the process of space launching, the system carries out simulation, field situation analysis of rocket launch, and flight-safety-targeted safety control, while emergency response would be made to save the launch vehicle. The structural rescue system integrating launch vehicles and the launch field offers emergency response preplan and measures, which could improve emergency support capabilities and provide real-time decision-making tools with great benefit.

Through researches of network-based and intelligent launch vehicle flight and flight emergency decision on GIS, the system combines simulation and spatial information processing for descriptions of the surroundings of the fault site and the range of fault, providing emergency response decision support and data-management technologies to solve safety issues concerning space launch and flight.

Emergency response decision refers to the emergency response to an emergency based on spatial information when an event occurs, which protects people's life and property and reduces losses, as shown in Figure 5.22.

Once any fault occurs, the system would extract spatial and property information like population, urban facilities, meteorology, terrain, environment of the accident location based on rocket trajectory data and related geographic information. According to the state information of the launch vehicle (size, fuel, and toxic or not), the system determines the explosion location and the range of shrapnel dispersion. The gas diffusion as well as their affected area in GIS could respond in a real-time and intuitive manner to estimate the losses and analyze the site situation in which integrating emergency database, preplan library, and emergency response model provide support for emergency decision.

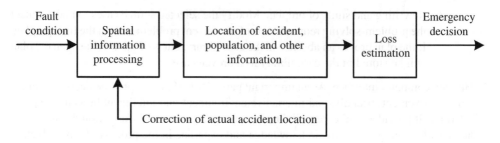

Figure 5.22   Emergency response based on spatial information.

5.5.3.1   Representation of Intelligent Emergency Response Decision

Preplan refers to decision plan targeting assumptions or "scenario" conditions made in advance, which is an implicit rule. The mechanism describes preplans by vector space methods and then represents the preplan based on an object-oriented framework.

The preplan is represented by ⟨*problem description and description explanation*⟩ or ⟨*problem description, description explanation, effect description*⟩. The preplan is described through main features (properties), the feature vector of which is as below:

Assume the preplan space is S, the problem space P, condition space T, decision space R, and preplan library CB. The preplan library CB is represented as $CB = \{cs_1, cs_2, \ldots, cs_k\}$, in which k is the number of real cases in the preplan library, $cs_i \in CB$.

Each preplan cs is composed of feature vector c and decision vector r: $cs_i = (c_i, r_i)$, in which $cs_i \in CB$, $cs_i \in T$, and $r_i \in R$. The latter two are the feature vector and decision result vector of preplan cs_i.

The problem descriptions of decision are divided into three types: feature property description, marked as a^e; theme property description, marked as a^s; and environment property description, marked as a^c.

Preplan feature vector c could be understood as the integration of a limited number of properties and the property values, that is,

$$c = \left\{ \left(a_i^e, v_i^e\right), \left(a_j^s, v_j^s\right), \left(a_k^c, v_k^c\right) \right\}$$
$$= \left\{ \left(a_1^e, v_1^e\right), \ldots, \left(a_l^e, v_l^e\right), \left(a_1^s, v_1^s\right), \ldots, \left(a_m^s, v_m^s\right), \left(a_1^c, v_1^c\right), \left(a_n^c, v_n^c\right) \right\} \tag{5.82}$$

where l, m, and n represent the number of feature property, theme property, and environment property of the preplan; v stands for the value of each property.

Assume the problem p shares a similarity of $\delta_i = f(cs_i, p)$ with preplan $cs_i \in CB$. For $q \in P$, by reflecting q in condition space T, we get problem feature vector $p \in T$. Match p with the condition vectors c_i of preplan $cs_i \in CB$, and we get similarity δ_i. Select the decision vector r_j of the highest similarity (such as δ_j) above the threshold which corresponds to preplan cs_j.

The preplan-based description in this section is composed of the following parts:

1. Category: Various emergencies that might occur in space launch and flight.
2. Decision condition: It refers to problem descriptions of the preplan, including feature condition, exclusive condition, and common condition. Feature condition refers to the

characteristic properties of the problem concerned, serving as feature descriptions of decision preplan, which is presented by key words. The exclusive condition refers to the theme property of problem properties, which is a special decision condition for the emergency response, such as the type of leaked poison gas, gas concentration, and physical and chemical properties of the gas. The common condition refers to the environment property of the problem properties, which is a description of the environment information in case of accident, including the flight period of the launch vehicle, longitude and altitude of the explosion, wind power, wind direction, etc.

3. Methods and measures: Take decisions according to accident category and decision-making. Define the weight according to the effects of each feature on decision's properties.

5.5.3.2 Calculations and Matching of Reasoning Similarity of Intelligent Emergency Response Decision

Preplan-based reasoning refers to the search of similar problems in the preplan library in order to get solutions of current problems. Therefore, after entering the target problem, you need to look for the most similar preplan to the decision conditions in the preplan library and policy conditions.

Measurement of Similarity
In the assessment of case similarity, it is necessary to establish a similarity calculation function in order to compare the current decision with the preplanning decision conditions. Set up the similarity function: sim: $U \times CB \rightarrow [0,1]$. U is the target domain, that is, the target preplans collection, and CB is the preplans collection in the preplan library. It represents the similarity of target preplan x to source preplan y with sim (x, y), in which $x \in U$ and $y \in CB$. Obviously, it has the following characteristics: $0 \le \text{sim}(x, y) \le 1$, $\text{sim}(x, x) = 1$, and $\text{sim}(x, y) = \text{sim}(y, x)$.

The role of problem properties contained in a preplan has different effects in the calculation of similarity. For this reason, it is necessary to weight different problem property with different values. Assume a preplan has a number of n problem properties, then $g_1 + g_2 + \cdots + g_j + \cdots + g_n = 1$, in which $0 \le g_j \le 1$, $j = 1, 2, \ldots, n$, g_j is the weight value of property j.

Assume the problem description of a preplan contains properties which total N in number n, and they are marked as $A_1, A_2, \ldots, A_n$ respectively. Their *co* domain is set as dom(A_1), dom(A_2), …, dom(A_n). Represent each property of decision problem T and preplan R in the preplan library with vector: $V_t = (a_{ti})$, $a_{ti} \in A_i$, $i = 1, 2, \ldots, l$; $V_r = (a_{rj})$, $a_{rj} \in A_j$, $j = 1, 2, \ldots, m$. Calculate the similarity between the two information objects:

$$\text{sim}(V_t, V_r) = \text{sim}(a_{ti}), (a_{rj}) = \text{sim}(a_{t1}, a_{r1}) \times g_1$$
$$+ \text{sim}(a_{t2}, a_{r2}) \times g_2 + \cdots + \text{sim}(a_{tn}, a_{rn}) \times g_n \tag{5.83}$$

In the formula, g_i ($i = 1, 2, \ldots, n$, $n = \min(l, m)$) represents the weight of each property. The most commonly used similarity calculation functions include the following types:

1. Tversky contrast matching function: a calculation approach based on a probability model;
2. Improved Tversky matching method: take into account different weights of various properties in the property collection in the two cases.

3. Distance measurement method or the nearest neighbor algorithm: calculate the similarities between two cases through the distance between two objects in the feature space. In CBR reasoning, most of the instance searches adopt the nearest neighbor algorithm.

   In addition, there are partial similarity technology and similarity calculation based on fuzzy set theory, etc. This system utilizes the distance measurement method.

Calculation by Distance Measurement Method
The distance measurement method or the nearest neighbor algorithm acquires the similarity between two preplans by calculating the distance between two targets in the feature space. In order to calculate the property similarity, it is necessary to define the codomain of each property to standardize the value, especially when the codomain belongs to symbol collection, and then quantifies the differences between property values.
   First introduce distance dist:

$$\text{dist}: \text{dom}(A_i) \times \text{dom}(A_j) \cdot 0 \leq \text{dist}(a_t, a_r) \leq 1$$

Adopt Minkowski distance measurement method, which is defined as:

$$\text{dist}(x, Y) = \sqrt[r]{\sum_{i=1}^{n} |X_i - Y_i|^r \times \omega_i^r} \tag{5.84}$$

If X_i and X_i represent the value of property i of preplan X and Y, r is the index and ω_i is the value of weight.
   Both distance and similarity could be used to describe the similarity between two preplans and their relation could be represented by:

$$\text{sim}(X, Y) = \frac{1}{1 + \text{dist}(X, Y)} \tag{5.85}$$

Calculation and Matching of Preplan Reasoning of Intelligent Emergency Response Decision
Because it is possible that the preplan doesn't match the actual problem completely, it is necessary to set a threshold. As long as the similarity between the two is greater than the threshold t, it is selected as the preplan candidate, that is, $\text{sim}_i > t, t \in (0, 1)$.
   As the properties of preplan problems are different, the corresponding weights also vary. Therefore, it is possible that the preplans share a low similarity as a whole but individual property enjoys a high degree of similarity with each other. Therefore, in the process of matching, the system would search out preplans that have a low overall similarity but share a high similarity in certain properties to provide more information for rewriting and with the decision makers. The flowchart is shown in Figure 5.23.

Realization of Preplan Reasoning in Intelligent Emergency Response Decision
The modifications and adjustment of preplan are very important in preplan reasoning. When there is no preplan that exactly matches with the problem to be solved, the system has to find preplan, which relatively resembles the problem to be solved, and then implement appropriate

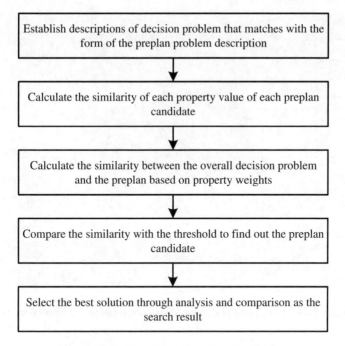

Figure 5.23   Flowchart of search and matching.

adjustments to enable it to adapt to the new situation. The correction technology can replace part or all of a decision plan with something else. The following two scenarios require plan amendments:

1. If the preplan searched through the nearest neighbor algorithm shows the highest degree of similarity with individual properties enjoying lower similarity, especially property with high weight value, it is suggested to analyze preplan that demonstrates lower overall similarity but higher property similarity, followed by extraction of correspondent decision content and replacement of the content that needs to be corrected.
2. If the preplan library fails to find the preplan that meet the requirements, the system would search relevant case knowledge and produce template based on related preplans to generate a new emergency response plan. The case library contains extensive factual knowledge, which is more comprehensive than information in the case library. Therefore, it can be used as a source of case modification knowledge.

5.5.3.3   Simulation Analysis and Result

In the process of space launch and flight, the system would integrate the trajectory data with geographic information visual simulation or spatial information processing based on the needs of launch vehicle safety control and emergency support to describe flight tracks, impact point and its range as well as spatial and property information like population, urban facilities, meteorology, terrain, environment. According to the state information of the

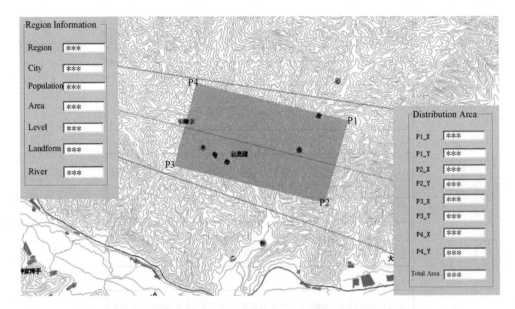

Figure 5.24   Simulation of shrapnel dispersion location and dispersion area.

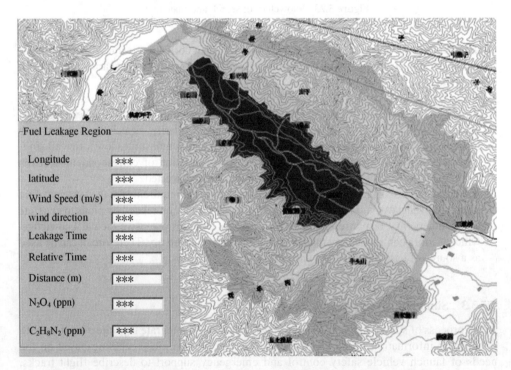

Figure 5.25   Simulation of fuel leakage and diffusion concentration contour.

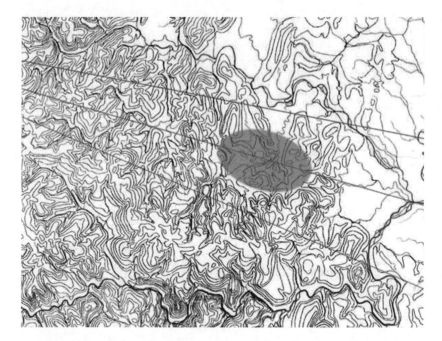

Figure 5.26   Simulation of impact point location prediction and affected area.

launch vehicle (size, fuel, and toxic or not), the system determines the explosion location and the range of shrapnel dispersion and gas diffusion as well as their affected area in GIS in a real-time and intuitive manner. Based on the emergency response database, preplan library, and safety emergency processing model, the system implements the simulation of flight safety of the launch vehicle and analysis of the site situation to find out the most similar preplan that matches with the decision conditions in the case library through the reasoning technology and offers support for emergency control processing and management.

Figure 5.24 uses shrapnel distribution area calculation model to simulate the scenario of an explosion of the launch vehicle, which describes the central point of exploded shrapnel and simulates the shrapnel dispersion area. The system offers the coordinates of the shrapnel central point and the dispersion area (the rectangle area).

Figure 5.25 simulates the scenario of rocket fault, in which measures are taken against fuel leakage. The simulation offers information, including the leaked fuel concentration contour, in which the curve boundary within each part of the map is the contour of lead fuel concentration.

Figure 5.26 simulates the predicted impact point of the launch vehicle and the affected area. The system offers the coordination of the impact point and the affected area (the ellipse area).

Based on the space launch features, the system combines intelligent decision support technology with GIS technology to offer descriptions of safety channel, impact point predictions. The flight parameter and flight tracks in the rocket flight process, and uses the shrapnel dispersion model and gas diffusion model under fault state to describe fragments dispersion range and fuel-affected area in a real-time and accurate way, achieving simulation of launch vehicle and flight safety, site situation analysis, and intelligent emergency controls strategy.

6

Development Tendency of Space Launch Test and Control

In the twenty-first century, the space and its resources will be fully exploited and utilized. Various industries will become increasingly dependent on space, and China will substantially increase the strength and diversity of its space launch. Development and utilization of space resources calls for low-cost, safe, fast capability of space exploration on a daily basis, which proposes higher requirements for spacecraft control and decision-making.

6.1 Technique and Methods of Space Launch Test and Control

The complexity of space launch control and decision-making involves techniques of integrated system analysis, test and control theories, and intelligence in combination with mathematics model. The latest control structure and method to solve feature analysis, diagnosis, model establishment, and control of complex objects in the process of space launches are reflected by information concerning key data from testing and launch as well as TT&C. Development of a theoretical system and technologies of space launch model, controlling and optimizing the decision-making and control mechanism of intelligent launch vehicle. The future launch test and control system will be larger, more complicated and reliable, which promote the rapid development of related theories, methods, and technologies. This phenomenon is mainly manifested in the following aspects:

1. New testing techniques and theories: Based on the characteristics of the future launch missions, researches will focus on an efficient and reliable test launch procedures, which adapt to the needs of different missions. In the future, flexible test technology will be the core of future directions of development. In addition, technologies such as remote network test and quick test in parallel also received extensive attention. In response to this

---

Intelligent Testing, Control and Decision-Making for Space Launch, First Edition. Yi Chai and Shangfu Li.
© 2015 National Defense Industry Press. Published 2015 by John Wiley & Sons Singapore Pte Ltd.

technological trend, researchers will focus on optimization of flexible testing scheme, parallel test scheduling, and data-driven testing data analysis and fusion in terms of theories and methods.

2. Technique and theory of fault diagnosis and prediction: In the future, the complex and varied missions of space launch will demand ensuring of highly reliable launch mission without any error, troubleshooting in which to meet the demands of high precision and immediate response. The fault diagnosis technology will focus on the fault diagnosis system based on simulation. Researches on fault diagnosis theory and techniques will mainly study small sample data-driven fault prognostics and system maintenance, quick analysis and location of complex system intelligent fault diagnosis, and other key issues.

3. Theory and technologies of real-time control and decision-making: Future space launch control and decision-making requires the system to respond more accurately, faster, safer, and more reliable, suitable for all kinds of space exploration missions and information technologies. The decision-making and control technology, based on simulation and intelligent expert system, is developing toward system monitoring technology, zero-window launching, or automatically restructure technology of a launch program. Study on the theories and related methods mainly focuses on rapid assessment of system performance, decision support based on test data and expert knowledge integration, and assessment and optimization of a launch program.

6.2   Informatization and Intellectualization of Space Launch Test and Control System

Space launch test and control system have moved from the manual control test, mechanical and electrical analog automation test and launch stage toward the stage of computerized automatic measurement and control. What is the direction of space launch test and control in the future? Has the computer automatic testing and control system met the requirements of future space test launch and control?

In recent years, spacecrafts such as satellites and deep-space probes are developing toward multi-functional, multi-redundancies, and high reliability. Development of fiber-optic and laser gyro navigation technology, civilian use of GPS system, and the control system of launch vehicle are also evolving toward platform +IMU strapdown +GPS/GLANASS/Beidou redundant system. Therefore, in the process of test launch, new requirements were put forward for test, analysis, and evaluation of the system. First, the dramatic increases of data in testing process will greatly diversify the types of data. Second, the data acquired during the testing process must go through separate evaluation as well as portfolio analysis, which complicates the diagnosis process. Third, it is necessary to optimize various data collections to choose the most suitable data and carry out relevant evaluations, which also makes the decision-making process more complex.

According to the development of launch vehicle and spacecraft, the new requirements that are put forward for test launch and control system can be summarized as follows: real-time data processing and visualization in various ways, online performance test and comprehensive diagnostic assessment, optimization and intelligent control, independent evaluation, and intelligent decision-making. The future space test launch and control system would be highly autonomous test and control system, featuring information and intelligence.

Therefore, the intelligent test and control of space launch will experience the following changes:

1. Evolving toward the advanced development: As we all know, things are getting improved day by day. So is the intelligent test and control system of space launch. Its development from lower to higher class is mainly reflected in the development of simple intelligence to sophisticated intelligence. At present, the intelligent testing is mainly used in multiple processing of testing data and intelligent evaluation of new testing data. The system has not acquired self-study capability while bearing limits within independent decision-making and intelligence, far from meeting the needs of space launches. Intelligent aerospace test launch is gradually developing from lower level to advanced level along with the technology and expertise.

2. Evolving from a single to integrated systems: Due to limitations of technical conditions and expertise, intelligent test launch applies to only certain subsystem of the aerospace test and control system. These applications enhance only the intelligent test and control capabilities of a single system or single subsystem, with the intelligent test and control system not working as a whole. It is necessary to carry out in-depth study of complex system modeling, intelligent algorithms, and system integration, achieving intelligent space test launch and control system.

3. The decision support system (DSS) develops toward autonomous decision-making: The introduction of intelligent method into the decision-making process provides decision makers with information on decision support in the first place. As to the adoption of specific decision, the system would not give a definitive answer but allows the decision maker to choose by himself according to the actual situation. Decision support system still needs improvement of information reasoning and evaluation. The autonomous decision-making function is one of our highest goals in designing the intelligent system. Because the space launch system enjoys stringent real-time, large-scale features with complicated constraints, multiple targets and various uncertainties, its development toward autonomous decision-making mechanism sensing, self learning, autonomous evaluation, and decision making capabilities, each of which has formed the focus and direction of intelligent researches.

4. Evolving from a single intelligent-algorithm toward integrated intelligent algorithm: For example, a simple control system can be transformed into an intelligent control system by a certain single intelligent algorithm to optimize the system's control performance. For complex systems, the single intelligent algorithms alone could hardly achieve satisfied performance of the system because each algorithm has its drawbacks and limitations. In order to address the issue of complex intelligent control and decision-making, the intelligent algorithm that combines two or more intelligent algorithms comes into being. The intelligent test and control of a space launch system involved multiple subsystems in various fields. To achieve a high level of intelligentization, a single intelligent algorithm is clearly not enough. Therefore, the integrated algorithm of intelligent test and control system for space launch is one of the focuses of future researches.

The intelligent test and control system of space launch in the future will inevitably be an integrated and complex system which is composed of multiple subsystems with a tridimensional structure. Researches on how to optimize theories and applications of intelligent test and control system of space launch will be the focus in the future.

The author believes that space test launch and control system in the future needs to be paid attention based on the following aspects:

1. The intelligent test and control have just been introduced into the space test launch and control system. Before this, they have been mainly applied in computer automated testing—although there is only a word of difference between the automated test and intelligence test (automated and intelligent are different in essence).
2. The current applications of intelligence methods are subject to subsystems in order to address the issue of local application of single issues without the capability of constituting a complete intelligent control system. Integration is one of the main issues to be tackled in the next step.
3. Intelligent systems suitable for test launch and control of aerospace: Researches on the application of intelligent methods in the field of space launch and intelligent algorithm suitable for space test launch still have a long way to go. This drawback will inevitably limit the development of an intelligent test system of aerospace. It can be said that algorithm of space intelligent test launch could preclude the development of space the intelligent test system.
4. The fundamental difference between the intelligent system and automated system lies in that the intelligent system has capabilities of independent learning, judgment, reasoning, and decision-making while automated system does not have this capability. The learning and reasoning function of the intelligent system is mainly based on rich domain knowledge collections. The capability of sorting, searching, and refining knowledge in the field of space launch plays a key role in promoting the development of space intelligent test launch and control system.

Bibliography

AlMutawa J, Identification of Errors-in-Variables Model with Observation Outliers Based on Minimum-Covariance-Determinant. American Control Conference, 2007. New York: IEEE. 2007, 134–139.

Atherton D P, Bather J A, Briggs A J, Data Fusion for Several Kalman Filters Tracking a Single Target, IEE Proceedings-Radar, Sonar and Navigation, 2005, 152(5): 372–376.

Bao P, Zhang H, Noise Reduction for Magnetic Resonance Images via Adaptive Multiscale Products Shareholding, Medical Imaging, IEEE Transactions on, 2003, 22(9): 1089–1099.

Bardina J, Rajkumar T, Intelligent Launch and Range Operations Virtual Test Bed (ILRO-VTB). Proceedings of SPIE—The International Society for Optical Engineering. Orlando, FL, United states: SPIE, 2003, 5091: 141–148.

Bardina J, Thirumalainambi R, Modeling and Simulation of Shuttle Launch and Range Operations. ESM 2004: 18th European Simulation Multi Conference. Moffett Field, CA, United States: NASA Ames Research Center, 2004: 13–16.

Bardina J E, Thirumalainambi R, Distributed Web-Based Expert System for Launch Operations. Proceedings—Winter Simulation Conference. New York: Institute of Electrical and Electronics Engineers Inc, 2005: 1291–1297.

Blair W D, Rice T R, A Synchronous Data Fusion for Target Tracking with a Multitasking Radar and Option Sensor, SPIE, 1991, 1482: 234–245.

Bossé É, Roy J, Paradis S, Modeling and Simulation in Support of the Design of a Data Fusion System, Information Fusion, 2000, 1(2): 77–87.

Cai Y-W, Yu G-H, Research on Test Method for Launch Vehicle, Journal of the Academy of Equipment Command & Technology, 2005, 16(4): 61–65.

Carlson N A, Federated Filter for Multiplatform Track Fusion. Proceedings of SPIE—The International Society for Optical Engineering. Bellingham, WA, United States: SPIE, 1999, 3809: 320–331.

Cesar B C F J, Waldmann J, Covariance Intersection-based Sensor Fusion for Sounding Rocket Tracking and Impact Area Prediction, Control Engineering Practice, 2007, 15(4): 389–409.

Chai Y, Ling R, Zhang Z, Li S, Safety Control Decision Making Based on Multi-source Information Fusion for Satellite Launch and Flight. Chinese Control and Decision Conference, 2008, CCDC 2008. Yantai, Shandong: Inst. of Elec. and Elec. Eng. Computer Society, 2008: 3689–3692.

Chen W, Application of Multi-scale Principal Component Analysis and SVM to the Motor Fault Diagnosis, International Forum on Information Technology and Applications, China: IEEE Press, 2009: 131–134.

Chen H, Chang K C, Novel Nonlinear Filtering & Prediction Method for Maneuvering Target Tracking, Aerospace and Electronic Systems, IEEE Transactions on, 2009, 45(1): 237–249.

Chen X, Zheng Q, Guan X, Lin C, Study on Evaluation for Security Situation of Networked Systems, Journal of Xi'an Jiaotong University, 2004, 38(4): 405–408.

Cheng C Y, Hsu C C, Chen M C, Adaptive Kernel Principal Component Analysis (KPCA) for Monitoring Small Disturbances of Nonlinear Processes, Industrial & Engineering Chemistry Research, 2010, 49(5): 2254–2262.

Cho H W, Nonlinear Feature Extraction and Classification of Multivariate Process Data in Kernel Feature Space, Expert Systems with Applications, 2007, 32(2): 534–542.

Cho H W, An Orthogonally Filtered Tree Classifier Based on Nonlinear Kernel-Based Optimal Representation of Data, Expert Systems with Applications, 2008, 34(2): 1028–1037.

Cockrell C E, Davis S R, Robinson K, Tuma M L, Sullivan G, NASA Crew Launch Vehicle Flight Test Options, Acta Astronautica, 2007, 61(1–6): 438–449.

Cook S, Hueter U, NASA's Integrated Space Transportation Plan-3rd Generation Reusable Launch Vehicle Technology Update, Acta Astronautica, 2003, 53(4–10): 719–728.

Cook S, Morris J C E K, Tyson R W, Technology Innovations from NASA's Next Generation Launch Technology Program. International Astronautical Federation-55th International Astronautical Congress, 2004, 13: 8401–8411.

Deng Z, Optimal Estimation Theory with Applications—Modeling, Filtering and Information Fusion Estimation, Harbin: Harbin Institute of Technology Press, 2005.

Du C, Pan J, Integrated Design of Ground Test Launch and Control System for New Generation Launch Vehicle, Aerospace Control, 2004, 22(1): 55–57.

Du C, Pan J, Integrated Design of Ground Test Launch and Control System for New Generation Launch Vehicle, Aerospace Control, 2004, 22(2): 50–52.

Facco P, Bezzo F, Barolo M, Nearest-Neighbor Method for the Automatic Maintenance of Multivariate Statistical Soft Sensors in Batch Processing, Industrial & Engineering Chemistry Research, 2010, 49(5): 2336–2347.

Fan Y-P, Chen Y-P, Huang X-Y, Chai Y, Gao B, Research on the Leak Current Fault Diagnosis System for Launch Vehicle Control System, Journal of Astronautics, 2004, 25(5): 508–513.

Fan Y-P, Chen Yun P, Chai Y, Huang X-Y, Han F-Q, Gao B, Research on Processing Method of Singularity Detection and Noise Elimination Based on Wavelet Transform for Data Measured in Launch Vehicle Aviation, Journal of Astronautics, 2005, 26(5): 591–624.

Gao B, Han F, Huang X, Li B, Compare Research of Two Filter Approaches on the Data of Rocket, Journal of Chongqing University (Natural Science Edition), 2003, 26(1): 25–27.

Geng Z Q, Zhu Q X, Multi-scale Nonlinear Principal Component Analysis (NLPCA) and Its Application for Chemical Process Monitoring, Industrial & Engineering Chemistry Research, 2005, 44(10): 3585–3593.

Goharrizi A Y, Sepehri N, A Wavelet-Based Approach to Internal Seal Damage Diagnosis in Hydraulic Actuators, Industrial Electronics, IEEE Transactions on, 2010, 57(5): 1755–1763.

Guan T, Xiong H, Luo Z, A Kind of Interface Adapter for 1553B's Bus, Measurement & Control Technology, 2003, 22(9): 39–45.

Guber A L, Application of GIS for Cassini Launch Support. U.S., NTIS No: DE2001-12526/XAB, 1993. Available: http://www.osti.gov/scitech/servlets/purl/12526. Accessed May 27, 2015.

Hall D L, Llinas J, An Introduction to Multisensor Data Fusion, Proceedings of the IEEE, 1997, 85(1): 6–23.

Han C, Zhu H, Multi-sensor Information Fusion and Automation, Acta Automatica Sinica, 2002, 28: 117–124.

Han C, Zhu H, Multi-source Information Fusion, Beijing: Tsinghua University Press, 2006.

Han K-Y, Lee S-W, Lim J-S, Sung K-M, Channel Estimation for OFDM with Fast Fading Channels by Modified Kalman Filter, Consumer Electronics, IEEE Transactions on, 2004, 50(2): 443–449.

Hardin D P, Marasovich J A, Biorthogonal Multi-wavelet on [−1,1], Applied and Computational Harmonic Analysis, 1999, 7(1): 34–53.

He Y, Wang G, Lu D, Peng Y, Multisensor Information Fusion with Applications, Beijing: Electronic Industry Press, 2007.

He W, Jiang Z-N, Feng K, Bearing Fault Detection Based on Optimal Wavelet Filter and Sparse Code Shrinkage, Measurement: Journal of the International Measurement Confederation, 2009, 42(7): 1092–1102.

Hu S, Huang L, Analysis and Diagnosis of Faults in Spaceflight Engineering, Basic Automation, 2003, 10(4): 296–298.

Hu C-H, Liu B-J, Sneak Circuit Analysis Based on Novel Coadjacent Neural Network Model for Reliability Control of Complex System, Acta Automatica Sinica, 2008, 34(2): 190–194.

Hu Y-Q, Chai Y, Li P-H, Real-Time Monitoring for Multivariate Statistical Process with Online Multiscale Filtering, Journal of Chongqing University, 2010, 33(6): 128–133.

Julier S J, Uhlmann J K, Unscented Filtering and Nonlinear Estimation, Proceedings of the IEEE, 2004, 92(3): 401–422.

Lee T S, Theory and Application of Adaptive Fading Memory Kalman Filters, Circuits and Systems, IEEE Transactions on, 1988, 35(4): 474–477.

Lee D S, Vanrolleghem P A, Adaptive Consensus Principal Component Analysis for On-line Batch Process Monitoring, Environmental Monitoring and Assessment, 2004, 92(1–3): 119–135.

Lee D S, Park J M, Vanrolleghem P A, Adaptive Multi-scale Principal Component Analysis for On-line Monitoring of a Sequencing Batch Reactor, Journal of Biotechnology, 2005, 116(2): 195–210.

Li Y, Cai Y-W, Research of Test Technique Based on 1553B Data Bus for New Generation of Launch Vehicle, Computer Automated Measurement & Control, 2005, 13(9): 964–966.

Li S, Chai Y, Huang X, Research on the Safety Controlling and Emergency System for Satellite Launching. Proceeding of the International Conference on Sensing, Computing and Automation. Chongqing: Watam Press, 2006: 3757–3760.

Li S-F, Huang X-Y, Che Z-M, Zhou M, The Emergency Decision-Making System for Space Vehicle Launching based on GIS, Journal of Astronautics, 2010, 31(4): 1200–1205.

Li S-F, Huang X-Y, Wei H-B, The Realization of Intelligent Decision-Making for Space Vehicle Launching and Safety Controlling, Journal of Astronautics, 2010, 31(3): 862–867.

Liu L, Post-flight Data Processing of Trajectory Measurement, Beijing: National Defence Industry Press, 2000.

Liu B-J, Hu C-H, Sneak Circuit Analysis Based on Neural Network, Journal of Astronautics, 2006, 27(3): 475–477.

Ma Y, Xie P, Chen Y, Test Systems for Launch Vehicle Based on United Test Tactic, Aerospace Control, 2004, 22(4): 88–90.

Mallat S, Hwang W L, Singularity Detection and Processing with Wavelets, Information Theory, IEEE Transactions on, 1992, 38(2): 617–643.

Mao W-B, Li S-F, Methods of Electrical Leakage Diagnosis for Launch Vehicle Based on Graph Theory Model, Journal of Astronautic, 2006, 27(sup): 166–169.

Nelson M, Mason K, General Principles for Data Fusion Systems. Proceedings of the Australian Data Fusion Symposium. Adelaide, Australia: IEEE, 1996: 223–228.

Pan Q, Yang F, Ye L, Liang Y, Cheng Y, Survey of a Kind of Nonlinear Filters—UKF, Control and Decision, 2005, 20(5): 481–489.

Pawlak Z, Rough Sets, Rough Relations and Rough Functions, Fundamental Informatics, 1996, 27(2, 3): 103–108.

Pawlak Z, Rough Sets and Intelligent Data Analysis, Information Sciences, 2002, 147 (1): 1212–217.

Qi F, Manned Spacecraft Technology, Beijing: National Defence Industry Press, 2003.

Rabelo L, Sepulveda J, Compton J, Moraga R, Turner R, Disaster and Prevention Management for the NASA Shuttle during Lift-off, Disaster Prevention and Management: An International Journal, 2006, 15(2): 262–274.

Rabelo L C, Sepulveda J, Compton J, Turner R, Simulation of Range Safety for the NASA Space Shuttle, Aircraft Engineering and Aerospace Technology, 2006, 78(2): 98–106.

Ren J, Cai Y, Application of New-Generation Test System Bus LXI in Aerospace Test, Aerospace Control, 2007, 25(5): 79–83.

Ren Y, Liu W, Zhang W, Measurement and Controlling System of Ground Hot-Firing Test for Rocket Roll-Control Engine, Binggong Xuebao/Acta Armamentaria, 2007, 28(2): 246–248.

Robinson M J, Determination of Allowable Hydrogen Permeation Rates for Launch Vehicle Propellant Tanks, Journal of Spacecraft and Rockets, 2012, 45(1): 82–89.

Romagnoli J A, Wang D, Robust Multi-scale Principal Components Analysis with Applications to Process Monitoring, Journal of Process Control, 2005, 15(8): 869–882.

Sala-Diakanda S N, Sepulveda J A, Rabelo L C, A Methodology for Realistic Space Launch Risk Estimation Using Information-Fusion-Based Metric, Information Fusion, 2010, 11(4): 365–373.

Salerno J, Information Fusion: A High-Level Architecture Overview. Proceedings of the 5th International Conference on Information Fusion, FUSION 2002. Annapolis, MD, United States: IEEE Computer Society, 2002, 1: 680–686.

Schlank J, Ground Testing, Aerospace America, 2003, 41(12): 70–75.

Shen R, Some Thoughts of Chinese Integrated Space-Ground Network System, Engineering Science, 2006, 8(10): 19–30.

Shen R, Zhao J, Development Tendency and Strategy of Space TT&C Technology of Our Country, Journal of Astronautics, 2001, 22(3): 1–5.

Steinberg A N, Data Fusion System Engineering, IEEE Aerospace and Electronic Systems Magazine, 2001, 16(6): 7–14.

Stenger B, Mendonça P R S, Cipolla R, Model-based 3D Tracking of an Articulated Hand. Proceedings of the IEEE Computer Society Conference on Computer Vision and Pattern Recognition. Kauai, HI, United States: Institute of Electrical and Electronics Engineers Computer Society, 2001, 2: II310–II315.

Sun S-L, Cui P-Y, Multi-sensor Optimal Information fusion Steady-state Kalman Filter Weighted by Scalars, Control and Decision, 2004, 19(2): 208–211.

Sun T, Neuvo Y, Detail-Preserving Median Based Filters in Image Processing, Pattern Recognition Letters, 1994, 15(4): 341–347.

Tong J, Cai Y, Application of LXI Bus in Test System for Launch Vehicles, Missiles & Space Vehicles, 2009, 29(2): 45–47.

Van Der Merwe R, Wan E A, The Square-Root Unscented Kalman Filter for State and Parameter-Estimation, Salt Lake, UT, United States: IEEE, 2001, 6: 3461–3464.

Wan E A, Van Der Merwe R, The Unscented Kalman Filter for Nonlinear Estimation. Adaptive Systems for Signal Processing, Communications, and Control Symposium 2000. Lake Louise, Alta.: AS-SPCC. The IEEE 2000. IEEE, 2000: 153–158.

Wang X, Challa S, Evans R, Gating Techniques for Maneuvering Target Tracking in Clutter, Aerospace and Electronic Systems, IEEE Transactions on, 2002, 38(3): 1087–1097.

Wang X, Kruger U, Irwin G W, Process Fault Diagnosis Using Recursive Multivariate Statistical Process Control. Proceedings of 16th IFAC World Congress. Prague, Czech Republic: The IFAC, 2005, 16: 287-292.

Wu J, Zhang Y, Chen Q, Transient Performance Simulation of a Large Liquid Rocket Engine under Fault Conditions, Journal of Aerospace Power, 1994, 9(4): 362–365.

Xu Y, Weaver J B, Healy Jr D M, Lu J, Wavelet Transform Domain Filters: A Spatially Selective Noise Filtration Technique, Image Processing, IEEE Transactions on, 1994, 3(6): 747–758.

Xu X H, Xiao F, Wang S W, Enhanced Chiller Sensor Fault Detection, Diagnosis and Estimation Using Wavelet Analysis and Principal Component Analysis Methods, Applied Thermal Engineering, 2008, 28(2–3): 226–237.

Yuan X, Xu H, Chen S, An Optimized Feature Selection Strategy in Liquid Rocket Engine Fault Diagnosis, Journal of Rocket Propulsion, 2008, 34(6): 2–5.

Zhang L, Ground Test Launch Control System of LM-2F Launch Vehicle, Missiles and Space Vehicles, 2004, 45(1): 34–37.

Zhang H, Study on Liquid Rocket Engine Fault Detection and Diagnostic Technology, Journal of Rocket Propulsion, 2004, 30(5): 41–45.

Zhang W, Wang J, Liu Y, Sun Q, Reliability Technology of the Ground Escape and Safety Control System in the Manned Space Project, Journal of the Academy of Equipment Command & Technology, 2006, 17(5): 107–110.

Zhou D-H, Hu Y-Y, Fault Diagnosis Techniques for Dynamic Systems, Acta Automatica Sinica, 2009, 35(6): 748–758.

Zhou D, Sun Y, Detection and Diagnosis Technology for Control System, Beijing: Tsinghua University Press, 1994.

Zhou D, Ye Y, Modern Fault Diagnosis and Fault-Tolerant Control, Beijing: Tsinghua University Press, 2000.

Zhou F-N, Wen C-L, Tang T-H, Chen Z-G, DCA based Multiple Faults Diagnosis Method, Acta Automatica Sinica, 2009, 35(7): 971–982.

Zhu Z-M, Qiu H-X, Li J-S, Huang Y-X, Identification and Elimination of Outliers in Dynamic Measurement Data, Systems Engineering and Electronics, 2004, 26(2): 147–149.

Index

absolute domain, 93
accelerometer, 5, 13, 67
access control, 54, 57–60, 62
access layer, 51–5, 57, 59
ACL, 61
activation function, 166, 177
active defense, 61
adaptive principal component analysis, 121, 122
advanced control, 8, 14, 15
air-to-air missiles, 3
analog circuit, 27, 174, 181, 186, 189
analog quantity, 15, 48
analytical model, 115, 118, 120
ant algorithm, 157, 158
ant colony algorithm, 113, 120, 150, 151, 156, 157
antiship missile, 3
antivirus system, 60
Apollo moon program, 159
Ariane, 13, 115
ARMA model, 70
artificial intelligence, 8, 14, 15, 22, 65, 116, 120, 121, 231
artificial satellite, 2, 4
assistant decision-making, 11, 17, 21
associated threshold value, 109, 110
association analysis, 75
associative memory, 174–6, 179, 185–7
astronaut, 2
astronautical engineering, 1, 2
astronautics, 2
asynchronous event, 34

attitude control, 6, 42, 66, 67, 124
attitude signal, 5
attractor, 145
attribute reduction, 119
attribute set, 92, 94
authentication, 24, 54, 58, 59, 61, 62
aviation zone, 67
azimuth reference, 68

backplane bus, 45, 47
ballistic missile, 2, 3, 116, 117
bandwidth limitation, 54
Beidou, 4
binary signal, 15
blow-off cylinder, 96
board level, 26
boundary domain, 93, 94
BP network, 160, 173, 174
bus topology, 27

C^3I, 19
C^4I, 13
cable swing beam, 21, 48, 50
CAMAC, 23, 25, 43, 44
CAN, 27
cardinal number, 94
cardinality, 94
carrier rocket, 19, 21, 23, 25, 29, 30, 37, 42, 43, 47, 59, 65–9, 98, 99, 114, 116, 167, 223, 238, 241
central-limit theorem, 77
certificate authority, 58, 61, 62, 77

Intelligent Testing, Control and Decision-Making for Space Launch, First Edition. Yi Chai and Shangfu Li.
© 2015 National Defense Industry Press. Published 2015 by John Wiley & Sons Singapore Pte Ltd.